Prepublication Reviews

"A switch to Application Specific Instruction Processors (ASIPs) from current ASIC Design is happening. Flexibility, product future proofing and the cost and complexity of deep submicron silicon design are the key contributors driving the change. The ability to offer software upgrades and keep products in the field longer is a business necessity. Designing and using ASIPs effectively is the new challenge. This book provides thoughts on the forces driving the change to ASIPs and provides insights into the necessary micro-architectural tradeoffs ASIP users must understand in order to evaluate their effectiveness in their application. Further, an intriguing software/hardware design approach is submitted for developers of ASIPs to consider. A detailed case study is employed to examine their programming language proposal which integrates an application component library. The target of this language case study is a Network Processor ASIP, the Intel® IXP1200. Although this is a first generation ASIP the power of the approach is intriguing. This research is necessary and of deep significance, and makes for a very interesting read."

MATTHEW ADILETTA
Intel Fellow, Director Communication Processor Architecture, Intel.

"This book makes a powerful case for ASIPs as a key building block of the next discontinuity in chip design, but it isn't just theory. It details a tested methodology for building ASIPs and provides examples from commercial tools."

RICHARD GOERING
Editorial Director, EDA, EE Times.

"The book provides a clear and validated design methodology, consisting of five main elements which are lucidly described and illustrated with great practical examples, to help designers quickly ascend the learning curve in becoming proficient with ASIP design."

AURANGZEB KHAN
Corporate Vice President, Cadence Design Systems.

"The performance, size, and complexity of modern Graphic Processor Units (GPU) are increasing at an explosive rate. Today, the design and verification of an ASIP like NVIDIA's GeForce 6800 Ultra is a formidable task even using our best methodologies. It's a real concern that tomorrow's GPUs may very well exhaust the ability of most traditional development flows to keep up. This book takes a real step toward turning the art of complex chip design into a craft. It details a disciplined, scalable development process that will scale with ever increasing scope and complexity of modern devices. I would encourage ASIP designers and software developers to read and consider the strategies presented in this book."

CHRIS MALACHOWSKY
Co-Founder/Fellow/Executive VP of Engineering, NVIDIA Corporation.

"In our technology business, major fundamental shifts occur once a decade. This new book describes the giant quake that's occurring in the world of electronics and semiconductors. Ready or not, the billions of consumers' lives are being changed by the pervasive use of heterogeneous embedded processors, known as ASIPs. The MESCAL group has looked into his crystal ball and sees clearly what companies must do to harness this disruptive technology. This book is a must read for anyone who wants to survive the coming processor revolution and its aftershocks."

ALAN NAUMANN
CEO, CoWare, Inc.

"Simplifying and automating the design of complex programmable platforms represents one of the grand challenges of modern electronic design. Without a systematic, abstract, but practical set of tools, building blocks and methodologies, the design of complex systems-on-chip will become technically painful and economically unsustainable. MESCAL represents a comprehensive effort to weave together the threads of application-specific processors, abstract algorithm development and practical systems modeling into a useful methodology and tool environment. This research should be understood by leading practitioners looking towards the next decade of innovation in embedded electronics."

CHRIS ROWEN
Founder and CEO, Tensilica, Inc.

"This book examines how to leverage the benefit of Moore's Law while addressing the associated design challenges. It provides a carefully balanced study of the practical use of ASIPs together with a forward looking exploration of the concept of programmable platforms. As a result this book will be a useful guide for architects, managers and engineers of complex systems. The systematic treatment of the ASIP design process with a focus on methodology makes this book unique and first of its kind."

ALBERT WANG
Co-Founder and CTO, Stretch, Inc.

"Methodologies are the key to design productivity and quality. This text is essential reading for those who want to maintain a competitive edge in system on chip design."

NEIL WESTE
Author of CMOS VLSI Design, Co-Founder of Radiata.

BUILDING ASIPS:
THE MESCAL METHODOLOGY

BUILDING ASIPS:
THE MESCAL METHODOLOGY

Edited by

MATTHIAS GRIES and KURT KEUTZER
University of California at Berkeley
Electronics Research Laboratory

 Springer

Library of Congress Cataloging-in-Publication Data

A C.I.P. Catalogue record for this book is available
from the Library of Congress.

ISBN-10: 0-387-26057-9   ISBN-10: 0-387-26128-1 (e-book)
ISBN-13: 9780387260570   ISBN-13: 9780387261287

Printed on acid-free paper.

Printed in the United States of America.

9 8 7 6 5 4 3 2 1   SPIN 11055181

springeronline.com

Contents

Contents

List of Figures

List of Tables

List of Definitions

Preface

When I was at AT&T Bell Laboratories, an integrated circuit designer, Mark Vancura, once told me: "A good talk is one where I go back to my office and do something differently." Over time this has also become my touchstone for quality research in electronic design automation and integrated circuit design. Application specific instruction-set processors (ASIPs) are as old as computing itself. Techniques and tools for the design of ASIPs have been an active research area for fifteen years now. The dream of the automatic generation of hardware and software – including simulators, assemblers, and compilers – has been commercially realized. The utility of these tools in enabling the exploration of the architectural design space has been demonstrated. More ASIPs are designed every year and ASIPs have become the building block of the highest performance programmable platforms of our era. For all this, the design of ASIPs is typically a chaotic process, and the commercial success of the result of this process is unpredictable. Working with our colleagues at CoWare and Tensilica we have developed a methodology that we hope will bring some rigor to ASIP design. In short, if you're working with ASIPs take this book home and read it. We hope that when you head back to your office you'll end up doing something differently.

KURT KEUTZER

Acknowledgments

Professional Acknowledgments

The work described in this book is the result of a long series of formal and informal collaborations. Pierre Paulin first introduced Kurt Keutzer to the notion of ASIPs on a visit to Synopsys in 1992. Even at that early date Pierre demonstrated to us that efficiently describing and evaluating architectures is the key to enabling efficient design space exploration. Pierre's visit lead directly to the SPAM (Synopsys Princeton Aachen MIT) work on tools for DSPs. This collaboration involved, among others, Kurt Keutzer, Stan Liao, and Steve Tjiang of Synopsys; Sharad Malik, Ashok Sudarsanam, and Guido Araujo of Princeton; Heinrich Meyr and Vojin Zivojnovic of Aachen; and Srinivas Devadas, George Hadjiyiannis, and Sylvina Hanono of MIT.

MESCAL was initiated with Kurt Keutzer's move from Synopsys to Berkeley. With MESCAL our interests broadened beyond DSPs to include benchmarking, software tools, and architectural exploration for ASIPs. The early MESCAL research effort included David August and Sharad Malik of Princeton. David went on to produce the Liberty simulation framework, and in time Sharad's interests shifted to applying formal approaches in runtime verification.

We were very fortunate to have excellent industrial contacts throughout our research. Our Intel liaison, Desmond Kirkpatrick, introduced us to Matthew Adiletta, Chief Architect of the IXP series. As detailed in the Introduction of this book, Matthew's visit to Berkeley greatly influenced our research directions. Christian Sauer, Research Engineer from Infineon Technologies, was an active and productive member of our group for three full years. There's no simple way to capture the value of having an industrially experienced integrated circuit designer as a daily collaborator in a sentence or two; Christian raised the quality of our work on every problem we approached. Kees Vissers spent a day a week with the MESCAL group for nearly two years. Kees first started working with us when he was Principal Architect at Trimedia and then continued after he became CTO at Chameleon. Kees always brought a fresh perspective to the group and we especially owe to him our understanding and appreciation of the importance of Lee's work on models of computation. Periodic interactions with Earl Killian, Monica Lam, Chris Rowen, and Albert Wang at Tensilica Inc. always stimulated our thinking about the potential for ASIPs. With Albert Wang's founding of Stretch Inc. we gained a new collaborator in Stretch's co-founder Gary Banta. Gary's thoughts on programming models for ASIPs have had a lasting impact on us. With the acquisition of Lisatek by CoWare, we began regular interactions with Alan Naumann who gave us a deeper appreciation for the commercial potential of ASIPs and their software tools.

Three names stand out by omission: Hugo DeMan, Masaharu Imai, and Peter Marwedel. Hopefully the future will give us more opportunities for collaboration with these three individuals who have contributed so much to ASIP-related research.

The MESCAL research effort is part of the Gigascale Systems Research Center (GSRC) and is supported, in part, by the Microelectronics Advanced Research Consortium (MARCO). We would especially like to thank Dan Radack and Simon Thomas for their personal interest and support of our work. The Tipi ASIP design framework is funded by Infineon Technologies. We would especially like to thank Dr. Ulrich Ramacher's technical interest and support of our work.

Thanks to Matthew Adiletta, Richard Goering, Aurangzeb Khan, Chris Malachowsky, Alan Naumann, Chris Rowen, Albert Wang, and Neil Weste for their positive feedback on the book.

Thanks to Daniel MacLeod, Lorie Brofferio, and Jennifer Stone; they help us keep the MESCAL team together in so many ways.

Heinrich Meyr, Rainer Leupers, Gerd Ascheid would like to acknowledge the key contributions of their Ph.D. students:

Gunnar Braun, Anupam Chattopadhyay, Malte Dörper, Tilman Glökler, Thorsten Grötker, Andreas Hoffmann, Manuel Hohenauer, David Kammler, Kingshuk Karuri, Torsten Kempf, Tim Kogel, Stefan Kraemer, Olaf L,thje, Achim Nohl, Stefan Pees, Andreas Ropers, Hanno Scharwächter, Oliver Schliebusch, Oliver Wahlen, Andreas Wieferink, Martin Witte, Jianjiang Zeng, Vojin Zivojnovic. Without their dedication, their hard work, and enthusiasm LISA would not exist.

Personal Acknowledgments

Kurt Keutzer would like to thank Barbara Creech for her patience and support.

Steven Leibson would like to acknowledge his wife Pat and daughter Shaina for their continued support of his writing.

Grant Martin would like to acknowledge his wife, Margaret Steele, and his daughters, Jennifer and Fiona, for their patience and understanding.

Contributing Authors

Gunnar Braun received his Diploma degree in Electrical Engineering from RWTH Aachen University in 2000. Since then, he published several technical papers and articles, and received the Best Paper Award at the Design Automation Conference in 2002. Mr Braun is one of the architects of the LISATek™ technology. Today, he is employed as a Principal Engineer at CoWare.

Anupam Chattopadhyay received the Master of Engineering in Embedded Systems Design from the University of Lugano, Switzerland, in 2002 and is currently pursuing the Ph.D. degree from the Institute for Integrated Signal Processing Systems (ISS), RWTH Aachen University, Germany. His research interests include automatic implementation of processors with LISA, architecture optimization techniques and a tool flow for reconfigurable ASIPs.

Matthias Gries is a researcher at Infineon Technologies working on microarchitectures for network applications. He was a post-doctoral researcher with Prof. K. Keutzer at UC Berkeley from 2002 to 2004. He received his Ph.D. from ETH Zurich, Switzerland, in 2001 for his work on QoS network processors. His interests include methods for developing ASIPs, system-level design, and analysis of real-time embedded systems.

Manuel Hohenauer received the Diploma in Electrical Engineering from RWTH Aachen University in 2000 and is currently working towards the Ph.D. degree in Electrical Engineering at the same university. His research interests include retargetable code generation for embedded processors with main focus on machine description generation for retargetable compilers from architectural descriptions and retargetable code optimizations.

Yujia Jin received his B.S. degree in EECS at UC Berkeley in 1999. He is currently a Ph.D. student in the EE department at UC Berkeley. His research interests include memory architecture, design space exploration, and soft core processors on FPGA.

David Kammler received the Diploma degree in Electrical Engineering from the Institute for Integrated Signal Processing Systems, RWTH Aachen University, in 2003 where he is currently pursuing the Ph.D. degree. He is one of the architects of the RTL hardware generation from LISA focusing on integration of automatically generated processor features and the generation of memory and bus interfaces.

Kurt Keutzer received his B.S. degree in Mathematics from Maharishi International University in 1978 and his M.S. and Ph.D. degrees in Computer Science from Indiana University in 1981 and 1984 respectively. Kurt spent the first seven years of his career at AT&T Bell Laboratories and the next seven years at Synopsys, Inc. where he became Chief-Technical Officer and Senior Vice-President of Research. He then joined the Department of Electrical Engineering and Computer Science at University of California at Berkeley as a Professor. He has published five books, over one hundred refereed papers, and has been involved in the formation of several companies as both an investor and advisor.

Chidamber Kulkarni is a researcher in Xilinx Labs focusing on network processing. He was a post-doctoral fellow at UC Berkeley with Prof. Kurt Keutzer from 2001-2003 working on network processors. He was at the Belgium-based research center IMEC from 1997-2001, working on memory-centric design methodologies for embedded systems. Chidamber obtained a Ph.D. in electrical engineering from Katholieke Universiteit Leuven, Belgium in 2001.

Steve Leibson is the Technology Evangelist for Tensilica. He formerly served as Editor in Chief of the Microprocessor Report, EDN, and Embedded Developers Journal. He holds a B.S.E.E. from Case Western Reserve University and worked as a design engineer and engineering manager for leading-edge system-design companies including Hewlett-Packard and Cadnetix before becoming a journalist. Leibson is an IEEE Senior Member.

Olaf Lüthje received the Diploma degree in Electrical Engineering from RWTH Aachen University in 1997. In the following he worked there as a research assistant in the DSP tools group of the Institute for Integrated Signal Processing Systems (ISS). In 2003 he joined CoWare Inc. where he works in the development team of the LISATek^{TM} product family. In 2005 he received a Ph.D. degree.

Grant Martin is chief scientist at Tensilica, Inc., Santa Clara, California. He graduated with a Bachelor's and Master's of Mathematics from the University of Waterloo, Canada. He then worked at Burroughs in Scotland, BNR/Nortel in Canada, Cadence in San Jose, and has been with Tensilica for one year. His interests are in system-level design, SoC Design, and embedded systems.

Heinrich Meyr received his M.Sc. and Ph.D. from ETH Zurich, Switzerland. He spent over 12 years in various research and management positions in industry before accepting a professorship in Electrical Engineering at the RWTH Aachen University in 1977. He has worked extensively in the areas of communication theory, digital signal processing and CAD tools for system level design for the last thirty years. In 2001 he has co-founded LISATek Inc. In 2003 LISATek has been acquired by CoWare, an acknowledged leader in the area of system level design. At CoWare Dr. Meyr has accepted the position of Chief Scientist.

Andrew Mihal received his B.S.E.E. from Carnegie Mellon University in 1999. He is currently a Ph.D. candidate at UC Berkeley. His research interests include programming abstractions for embedded multiprocessors, and as a hobby, heuristic techniques for disguising seams in digital image mosaics.

Matthew Moskewicz is pursuing a Ph.D. in electrical engineering at the University of California, Berkeley. His research interests include satisfiability, formal modeling and languages for system design, and constraint based programming. Moskewicz has a BSE in electrical engineering from Princeton University. He recently joined CommandCAD, Inc.

William Plishker is presently working towards a Ph.D. in Electrical Engineering at the University of California, Berkeley. His research interests include design automation techniques for programmable embedded systems. In 2000, he received his B.S. degree in Computer Engineering from the Georgia Institute of Technology.

Kaushik Ravindran is a Ph.D. student in the EECS Department at the University of California, Berkeley. His research activities include reconfigurable computing and embedded system design. Ravindran has a B.S. in computer engineering from the Georgia Institute of Technology. He is a student member of IEEE.

Christian Sauer heads the networked systems group at Infineon Technologies Corporate Research in Munich. After graduating from Dresden University of Technology in 1996 he joined Siemens CT labs working on processors for image processing, multimedia, and communications. From 2001-2004 he visited Prof. Keutzer's MESCAL group at UC Berkeley. His research interests are in ASIPs, SoC design, and embedded applications.

Oliver Schliebusch received his Diploma degree in Electrical Engineering in 2000 from the RWTH Aachen University. Since then, he is working towards the Ph.D. degree in Electrical Engineering. In January 2004 he has been appointed as chief-engineer at the Institute for Integrated Signal Processing Systems. Mr. Schliebusch is one of the architects of the LISA Processor Design Platform and leads the research activities on automatic processor implementation with LISA.

Niraj Shah is a product marketing manager at Catalytic, Inc. in Palo Alto, Calif. He recently received a Ph.D. in electrical engineering and computer sciences from the University of California at Berkeley. His research focused on programming models for application-specific processors. Previously, Niraj was a venture partner at ITU Ventures, an early-stage venture capital firm investing in companies emerging from leading research institutions.

Mel Tsai received his M.S.E.E. in 2003 and is currently a 6th year Ph.D. candidate at UC Berkeley. Tsai has a B.S. in electrical engineering from Michigan State University. His dissertation work is called RouterVM, a novel configuration interface and architecture for next-generation network appliances.

Scott Weber is pursing a Ph.D. in electrical engineering at the University of California, Berkeley. His research interests include tools and methods for the design and implementation of programmable platforms. Weber has a B.S. in electrical and computer engineering and a B.S. in computer science, both from Carnegie Mellon University.

Andreas Wieferink received the Diploma degree in Electrical Engineering with honors from RWTH Aachen University in 2000. Since then, he is working toward the Ph.D. degree in Electrical Engineering at the same university. His current research interests include Multi-Processor System-on-Chips (MPSoC),

Retargetable Processor System Integration and Hardware/Software Cosimulation.

Ernst Martin Witte received the Diploma degree in Electrical Engineering in 2004 from the Institute for Integrated Signal Processing Systems, RWTH Aachen University, Germany, where he is currently pursuing the Ph.D. degree. Currently, his research focuses on architecture exploration and implementation of application specific processors based on LISA and an ADL based design flow aware of reconfigurable programmable architectures.

Trademarks

Throughout this book we make reference to various software and hardware products and items. The following trademarks and registered trademarks are referred to in this book and are the property of the following organizations:

ACE Associated Computer Experts bv.: CoSy

Advanced Micro Devices, Inc.: Athlon

Advanced RISC Machines Ltd.: AMBA, ARM7, ARM9E, StrongARM

Agere Systems, Inc.: PayloadPlus

Berkeley Design Technology, Inc.: BDTIMark2000, BDTIsimMark2000

Bluetooth Special Interest Group: Bluetooth

Broadcom Corp.: Calisto

Cadence Design Systems, Inc.: NC-Verilog, Verilog

CEVA, Inc.: CEVA-X

CoWare, Inc.: ConvergenSC, LISATek

Cypress Semiconductor Corp.: NoBL

Freescale Semiconductor, Inc.: C-Port, PowerQUICC, PowerPC

HyperTransport Technology Consortium: Hypertransport

IEEE Standards Association: 802, POSIX

Infineon Technologies AG: TriCore

Integrated Device Technology, Inc.: Zero Bus Turnaround, ZBT

Intel Corp.: Celeron, Intel XScale, Pentium, Pentium II, Pentium III, Pentium 4

International Business Machines, Inc.: CoreConnect, PowerPC, PowerNP

Lexra, Inc.: NetVortex

Linus Torvalds: Linux

LSI Logic Corp.: ZSP, ZSP400, ZSP500

Microsoft Corp.: Windows

OpenMP Architecture Review Board: OpenMP

Open SystemC Initiative (OSCI): SystemC

PCI Special Interest Group: PCI, PCI Express, PCI-X

Philips Electronics N.V.: Nexperia, TriMedia

QDR Consortium: Quad Data Rate, QDR

Rambus, Inc.: RDRAM

Sonics, Inc.: SiliconBackplane

Sparc International: SPARC, UltraSPARC

StarCore LLC: StarCore

Sun Microsystems, Inc: Java, Solaris

Synopsys, Inc.: Behavioral Compiler, Design Compiler, DesignWare, Physical Compiler

Tensilica, Inc.: FLIX, Xtensa, Xplorer, XPRES

Texas Instruments, Inc.: OMAP

The Open Group: POSIX, UNIX

Wind River Systems, Inc.: VxWorks

Xilinx, Inc.: Virtex-II Pro

Chapter 1

INTRODUCTION AND MOTIVATION

Kurt Keutzer
University of California at Berkeley
Electronics Research Laboratory

We define a *programmable platform* as a *platform* [126] consisting of an assemblage of programmable components together with an integral *programming model*. These programmable components can be either software programmable or electrically (e.g. FPGA) programmable. The programmable components are bound together with a common communication fabric, such as a *network-on-a-chip* [22, 216, 118] for communication. We define an *application-specific instruction-set processor* (ASIP [206]) as a software programmable processing element tailored for a particular application. There are a number of reasons why the design and deployment of programmable platforms consisting of assemblages of ASIPs is of growing interest: System designers are looking for computationally efficient and economical ways to provide system solutions. ASIPs, individually and collectively, are providing an attractive approach in a growing number of application areas including graphics, video, networking, and signal processing. These programmable devices provide designers a high-performance and energy efficient alternative to general-purpose processors. They also have significant advantages over application-specific integrated circuits (ASICs) that will be discussed below.

A second motivation for interest in programmable platforms and their ASIP components comes from semiconductor manufacturers. As Moore's Law continues to increase the number of transistors that can be economically put on a die, the question arises: What are we going to do with all these transistors? Or, from a business standpoint: What are we going to build to keep our fabs full? Building more-and-more deeply pipelined general-purpose processors on a die together with large amounts of memory in the form of caches is no longer an attractive way to use silicon. Such devices are very expensive to build and are very inefficient in their use of power. As design costs become unmanageable and power becomes the fundamental limiting factor in integrated circuit design, programmable platforms become attractive. The ASIP component is an energy-efficient unit that can be economically designed and systematically replicated over a die to form programmable platforms.

Perhaps the strongest motivation for programmable platforms comes from the increasingly prohibitive cost of building ASICs. Numerous factors are making it increasingly difficult and expensive to design and manufacture ASICs. This has started a significant move towards the use of application-specific standard parts (ASSPs) rather than ASICs. While ASICs and ASSPs share a similar design methodology, the key difference is that the non-recurring design costs of ASSPs can be amortized over multiple applications. The key to applying a single integrated circuit to multiple applications is programmability. Thus, for the integrated circuit designer, programmability enables a higher volume to amortize design and manufacturing costs. For the application implementer, programmability provides a lower risk and shorter time to market.

Periodically we hear phrases such as "silicon is free" or "cycles are free" but the fact is that integrated circuits typically gain or lose design wins based on relatively small differences in speed, cost, or power dissipation. The flexibility provided by the programmability of programmable platforms comes with a performance and power overhead. The ability for ASIP-based solutions to get design wins depends very much on their ability to provide an attractive point in the space of power-delay-cost-productivity trade-offs. If ASIP-based solutions fail to achieve attractive power-delay design points they may lose design wins to general-purpose processor solutions. If ASIP-based solutions are too difficult to program then they may lose their productivity advantage over ASIC solutions. In short, while ASIPs are strongly motivated by current design trends, for an ASIP to gain design wins requires execution excellence. This book is devoted to trying to capture the key elements of a methodology to create a successful ASIP that can then be used as an efficient building block for larger programmable platforms.

Over the last seven years the MESCAL[1] research group has experimented with implementing applications on several ASIP platforms. We have also done paper studies and analyses of dozens more. In addition, we have had the good fortune to talk to a number of chief architects and developers of ASIPs. Through this process we have arrived at a simple five-element methodology that embodies our understanding of the best practices to be employed in developing an ASIP.

The remainder of this chapter elaborates on the themes outlined above. We first review past design discontinuities in electronic design automation to motivate why the industry is poised for another design discontinuity. We review the current challenges in digital IC design. We then suggest that programmable

---

[1]MESCAL is an acronym for Modern Embedded Systems, Compilers, Architectures, and Languages.

platforms built of ASIP components enable the next design discontinuity. We survey empirical evidence that ASIPs are a growing segment of the semiconductor market. We point to some problems in current ASIP design practice and finally outline a five-element methodology that addresses these problems.

1. What is a Design Discontinuity?

1.1 A History of Design Discontinuities in EDA

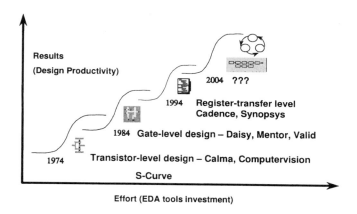

Figure 1.1. Design discontinuities in EDA.

Periodically designers abandon their existing tool sets and migrate to an entirely new design methodology. Such a migration is rarely done voluntarily. Learning a new approach is challenging and no designer is looking for more work to do. Nevertheless, as design deadlines approach, and existing tool-sets are not doing the job, designers are willing to try anything. Figure 1.1 shows how the evolution of design methodology adoption in the electronic design automation (EDA) tools industry follows a classical S-curve pattern. The Y-axis indicates the level of design productivity supplied by a design methodology such as register-transfer level synthesis. The X-axis indicates the amount of effort invested in the approach. Here we suggest R&D investment in the tool flow as a good metric for effort. The S-curve figure demonstrates how design discontinuities occur and design methodologies are typically adopted. At first a design methodology may undergo a period of significant investment before it shows any promise for improving designer productivity. In the second phase, as we ascend the "S" of the curve, there is typically a point of discontinuity at which only moderate incremental investment in the methodology leads to significant improvements in productivity. Finally as the "S" flattens out again significant investment in a tool methodology leads to negligible improvement in designer productivity. Figure 1.1 also shows that the EDA industry has seen a number of discontinuities so far. A brief review of the history of EDA should help us to understand where we stand today.

Graphic workstations improved the data-entry process for integrated circuits, but transistor-level design remained a mysterious and laborious art until Spice circuit simulation. First developed in 1973, Spice enabled designers to efficiently model a wide range of circuits. The combination of graphical entry and Spice simulation dramatically improved designer productivity of integrated circuits through the 1970's. These productivity improvements were impressive; however, EDA tool providers face a unique challenge: They must keep up with Moore's Law. Incremental productivity improvements that might be impressive in other industries fall far short of providing the improvements necessary to keep up with the 60% per year (or more accurately, 100% every eighteen months) increase in transistor density that the semiconductor industry has consistently provided.

By the 1980's productivity improvements through transistor-level design had flattened out, but in the meantime the same graphics stations and Spice tools laid the foundation for another design discontinuity. Using graphic workstations designers were able to build libraries of standard re-usable cells called "standard cells[2]." These cells were designed so that they could function reliably wherever they were placed on the circuit die. As a result they were large and, from the standpoint of many circuit designers, they reflected a very inefficient use of silicon. Nevertheless, extensive use of Spice simulation enabled tool providers to reduce the complex circuit properties of these cells to models consisting of a few tabular entries, and additional tool support for physical design evolved that was able to take graphs constructed from these cells, known as *netlists*, and automatically place them on a circuit die.

As circuits became too large to design by manually laying out transistor-by-transistor, the industry was poised for another design discontinuity. With well-characterized standard cells and automated place-and-route tools in place, the key missing tools were design entry and simulation at the gate level. The commercial solution came from Mentor Graphics (founded 1981), Valid Logic (1981), and Daisy Systems (1984). These companies provided tools that allowed designers to enter gate-level schematics rather than laying out transistors. Productivity was boosted not just because the gate-level schematics were at a higher level of abstraction; these gate-level schematics were free-floating logical structures, and did not require the logic-designer to engage in painstaking

---

[2]A parallel development occurred in the development of *gate-arrays* but for the sake of simplicity we focus only on standard-cell based design in this discussion.

physical design. Automated place-and-route tools would determine the precise physical layout of the cells and wires at a later step. Moving from transistor-level entry to gate-level entry together with automating physical layout were important steps toward improving designer productivity; however, the verification problem also needed to be addressed. Otherwise, relying on Spice simulation at the transistor level would simply create another bottleneck in design, and Amdahl's Law would indicate that design productivity would only be modestly improved. The key additional capability that logic-level tool suites provided was *logic-level* simulation. By moving the functional verification of the circuit from the transistor-level to the logic-level simulation speeds were improved by factors of 100-1000X.

Over the 1980's gate-level design, and in particular place-and-route, became more mature. By the mid-eighties if you sent a "sign-off" package consisting of a standard-cell library, a *netlist* of cells from the library, a set of simulation vectors, and a set of manufacture test vectors off to a semiconductor fabrication facility then you could be confident you would receive back integrated circuits that worked to your specification. As a result the logic-level netlist of cells became a new foundation on which higher-level design techniques could be built.

And that was a good thing: Because by the early 1990s gate-level design also ran out of steam. Designers found that manually entering 30-40,000 gates was simply too time consuming. Worse, to verify a system the entire gate-level design had to be entered. To address this issue hardware-description languages (HDLs) were developed that modeled circuits at the register-transfer-level (RTL). Designing at the register-transfer level designers needed only to describe the logical transfer functions between registers. The detailed logic implementation of those transfer functions need not be described. Semiconductor companies such as IBM, Intel, and Bell Laboratories all used HDLs that modeled circuits at the register-transfer level, but commercial use of HDLs began with Phil Moorby's Verilog language that appeared in 1985. Using Verilog, or other proprietary HDLs, designers again achieved a factor of 10X increase in simulation speed. This allowed designers to model and verify larger circuits. However, the implementation problem remained. After verifying a Verilog description of a circuit the design still had to be manually entered through schematic entry. First at IBM in the early 1980's, and then over the next decade at other vertical system companies such as Bell Laboratories, researchers demonstrated that gate-level netlists could be automatically generated from HDL descriptions of circuits described at the register-transfer level. Thus by building on top of the capability of automatic place-and-route tools developed in the gate-level schematic era, a new design methodology known variously as logic synthesis, HDL synthesis, or RTL synthesis, was born. By combining HDL-based entry and automated logic synthesis designers were able to generate gate-level netlists in a matter of weeks that would have required months of manual schematic entry. Designers experienced a 10X increase in productivity with a modest, in most cases, degradation in final circuit speed. During the 1990's HDL synthesis from the register-transfer level became the predominant mechanism of circuit design.

Before we ask the obvious question "So what's the next design discontinuity?" let us review the characteristics of the design discontinuities.

1.2   General Characteristics of Design Discontinuities

A design discontinuity has the following fundamental characteristics:

- There is a significant commercial push to achieve higher levels of productivity due to the increased number of transistors that can be economically placed on a semiconductor die – i.e. due to Moore's Law.

- The foundation of the design discontinuity is the creation of a base level, or *platform*, whose functional and physical properties are well understood. This platform must be able to be reliably and predictably manufactured.

- The "step up" in the new methodology comes from a *higher-level* abstraction that allows details of the foundation to be abstracted away.

- A methodology supporting a design discontinuity maps from the higher-level abstraction down to the reliable foundation. Elements of this methodology are:

   - A *design-entry* approach that offers 10X productivity improvement. The *holistic view* of the entered design provided by this approach *must be greater than the sum of its parts.*

   - A *functional verification* approach that offers 10-100X speed-up in verification.

   - An *implementation* approach that is predictable and reliable. This approach must offer "single-pass" success even if it loses as much as 10X in speed or area relative to existing methodologies on portions of the design. It is important that the new methodology provides "hooks" so that higher performance can be achieved when needed.

In addition, design discontinuities can be empirically observed to have the following characteristics:

- The move to a higher level of abstraction most often begins with

   - The development of a reliable foundation, then continues with

- *Design entry* and *verification/simulation* at a higher-level of abstraction and then

- Support for *implementation* by compilation or synthesis from the high-level of abstraction down to the foundation.

■ Since system developers are typically risk averse, a design discontinuity must be demonstrable with only a few designers. In other words, design discontinuities are typically first demonstrated by a small group of three to five designers finishing a design with a new methodology that a large group of 30 or more designers could not complete with the old methodology.

■ Design discontinuities are usually demonstrated by proprietary experimental tool suites used in vertically integrated system companies.

■ EDA tool vendors that were successful with the old methodology are slow and ineffectual at responding to the design discontinuity and often lose significant market share as a result.

■ Although methodologies embodying a design discontinuity are first demonstrated in large vertical system companies, new independent tool companies are typically the broad commercial suppliers of the new methodology. Transition from the proof-of-concept at the large system company (e.g. IBM's LSS) to broad commercial deployment by the independent commercial supplier (e.g. Synopsys' Design Compiler™) can take as long as a decade.

2. The Current Environment for Design

Having reviewed the basic elements of a design discontinuity let us examine the current environment of designing ICs. We will begin by looking at the *push* to a new methodology from the semiconductor industry as well as the *pull* to a new methodology from the emerging applications in the electronic systems industry.

2.1 The Push to a New Methodology from Semiconductors

2.1.1 Increasing Design Costs.
Designing an integrated circuit is getting increasingly expensive with each succeeding generation. Design difficulties arise from four distinct causes:

■ *Deep-Submicron Effects (DSM):* Designing in deep sub-micron geometries (< 250nm) raises a host of new challenges associated with the physical effects that begin to predominate at these small geometries. The primary change is the increase in interconnect delay as a fraction of the gate delay due to scaling effects. Since this is not available till physical design (place and route) is over, the traditional flow of logic synthesis, with simple interconnect wire-load models, followed by physical synthesis does not work anymore. This has resulted in multiple iterations with no promise of convergence. This is referred to as the *timing convergence problem* or more generally the *design convergence problem*. In addition to this, an increase in coupling capacitance results in crosstalk, thus compromising signal integrity. Other deep submicron effects include increased static power dissipation due to leakage. While design tools offer some support (at an increased expense) for both of the above issues, these problems are far from being solved.

- *Increased Complexity:* The flip side of smaller geometries is that we can now integrate more transistors on the same die. This is amplified by the fact that manufacturing advances have further increased possible die sizes (roughly a growth of 20% every four years). Thus, not only is each transistor harder to design, we have exponentially more of them to deal with. This complexity has put pressure on existing design tools; a different set of tools are required to better manage the hierarchical block level designs that ensue from this increased complexity.

- *Heterogeneous Integration:* Increased functionality of systems at lower costs requires the integration of heterogeneous functionality on the same die. In addition to the traditional digital part, it is not uncommon to integrate analog and mixed signal components on the same die. Further, independent of the case being made in this book, increasingly control and feature-support software is also part of the design. This heterogeneous integration requires diverse specialization and expertise in design groups.

- *Shrinking Time-to-Market:* While the above three factors arise out of technology challenges, a fourth factor arises from commercial challenges: The time-to-market for products is shrinking – this provides an added degree of difficulty in realizing commercially successful designs.

The above trend has been clear for several years now. The design productivity gap between what can be economically manufactured and what can be economically designed continues to grow today. The International Technology Roadmap for Semiconductors (ITRS) 2001 aptly summarized this: *"The main message in 2001 is this: Cost of design is the greatest threat to continuation of the semiconductor roadmap."* Subsequently, it goes on to state: *"In 2001, many design technology gaps are crises."* (The emphasis is original for both quotes.)

2.1.2   Increasing Manufacturing Costs.   Besides design costs, the non-recurring engineering (NRE) costs associated with manufacturing have also increased over the past several years.

- *Mask costs* for designs in today's 180-90nm technologies are in the 0.5-1M$ range. This is expected to only rise with further shrinking geometries for two reasons: The primary reason is that each mask layer costs more to produce in a smaller geometry. A second reason is that the number of mask layers increases due to adding wire layers or adding voltage thresholds.

- *Packaging costs* increase because smaller geometry designs have more individual components that in turn require more pins. Also, as the power density of chips increases higher-cost packages are necessary to dissipate the power.

- *Testing* 100M-1B transistor circuits at high operating frequencies poses significant challenges. We have reached the point where the cost to test a transistor is no longer negligible compared to the cost to manufacture it. ITRS 2001 states: *"Test costs have grown exponentially compared to manufacturing costs."* While this is largely reflected in a cost-per-part, there are non-recurring engineering (NRE) design costs associated with developing test strategies and test vectors.

The increases in NRE design and manufacturing costs have a direct impact on the nature and price of designed parts. For fixed volume parts, this results in possibly prohibitive per-part costs. If this cost is not acceptable, then it leaves the application designer no alternative but to implement the application in software on some existing programmable platform. For fixed cost parts, this requires a much larger volume to amortize these costs.

2.1.3   Cumulative Impact.   Research and development investment by the semiconductor industry is always high, but in any stable and profitable period of the semiconductor industry the amount of research-and-development investment that a company can apply to the development of integrated circuits will be fixed somewhere between 5% and 15% of total semiconductor product revenue. Worldwide semiconductor sales are highly cyclical. For example, revenues peaked in 2000 and fell off sharply in the years 2002-2003. Four years later, 2004 revenues came back on par to 2000.[3] Over its history the semiconductor industry has shown sustained double-digit growth. Goodall et al. observe an average 16% a year increase in revenues over 40 years of the

---

[3]http://www.icknowledge.com/economics/Status0305.html

industry [83]. This number is consistent with a 15.1% growth rate quoted in [104].

Formula 1:

$$design\_budget = semiconductor\_revenue \times \%\_of\_revenue\_for\_R\&D\text{-}budget$$

So, if the amount of money available to spend on integrated circuit design is as given in Formula 1, then, the above data is quite encouraging. It indicates that design budgets for integrated circuits can, adjusted for cyclical trends, steadily increase. If we want to do an estimate of the number of IC designs that we can afford to make, we can simply divide our budget by the cost of designing an integrated circuit to arrive at the Formula 2[4].

Formula 2: $number\_of\_ICs \leq \dfrac{design\_budget}{cost\_of\_designing\_an\_IC}$

Of course the cost of designing an IC varies according to a wide number of factors including: The IP included on the IC, the complexity of the design, software content, and so forth. Since we are primarily focusing on *trends*, we are interested in examining how design costs are *changing* with time. Precise published data on rising IC design costs has been hard to find; however, informal polling of design managers indicates that the total product cost of designing an integrated circuit has been increasing 50-100% per process generation from 180 nanometers through 130 nanometers to 90 nanometers. While it is natural to be suspicious of data on the costs of IC design presented by semiconductor vendors selling reconfigurable alternatives to ASICs, data predicting $30M product development costs in [193] appears to give accurate, perhaps even conservative, estimates of the costs of developing a significant system-on-a-chip (SoC) platform in 90 nanometers. In fact, informal reports of the product development costs of a SoC platform in 90 nanometers for 3G wireless applications can be over $50M. As another example, Sony, Toshiba, and IBM have jointly announced a $400M "investment" in the development of the Cell Processor, a joint development targeting the video gaming market. To understand the rising IC design costs let us break them down more in Formulas 3, 4, and 5.

Formula 3:

$$cost\_of\_designing\_an\_IC = design\_costs + initial\_manufacturing\_costs$$

---

[4]This line of formulation was spawned by presentations at DAC 2004 by Ivo Bolsens of Xilinx and Chris Rowen of Tensilica at a panel moderated by Kurt Keutzer. Basically we have plugged Rowen's formula on return-on-investment for semiconductors into Bolsen's concise analysis of the number of ICs that can be designed.

Formula 4:

$$design\_costs = staff\_years \times (loaded\_salary/year + tool\_costs/seat/year)$$

Formula 5:

$$initial\_manufacturing\_costs = (mask\_costs \times iterations) + (wafer\_lots \times iterations) + initial\_packaging\_costs + initial\_testing\_costs$$

Initial manufacturing costs include the costs of creating masks for the semiconductor processes, the cost of producing wafer lots to give chip samples, the costs of packaging dies from those wafers, and the cost of testing them. Section 2.1.2 gives a clear discussion of the rising costs in manufacturing. Packaging and testing costs primarily affect the per-unit cost of ICs, but taken as a whole initial manufacturing costs are growing significantly – perhaps doubling each process generation.

Even if initial manufacturing costs are doubling, it is the human design related costs that are most impacting the total product cost of designing an IC. The staffing for an IC grows and shrinks over the different stages of a design project, and tool usage varies as well. Formula 4 simply tries to approximate these effects by averaging them over the lifetime of a project. The key point is that because we continue to use a register-transfer level synthesis methodology, individual designer productivity is not improving. In fact, based on the discussion in Section 2.1.1, one might argue that design productivity per engineer is significantly decreasing. If this is not entirely the case it is only because of increased design re-use of IP blocks. Thus, as design complexity increases, staffing must increase *proportionately*. Because of the increasing complexity of the tool flow, the tool costs per seat (i.e. per engineer) are also increasing.

The reader may feel that this has been a long and protracted analysis to simply rationalize what is today ubiquitously observed: The number of IC designs is steadily decreasing. However, we hope to garner a few more insights from our analysis: Reflecting on Formulas 3, 4, and 5 in the light of Sections 2.1.1 and 2.1.2, it is easy to understand empirical data indicating 50 - 100% increases in total product development costs for ICs. Reviewing Formula 2, if the *design budget* is increasing at most 16% a year and *product development costs* increasing 50 - 100% per year, we can then expect the number of ICs designed in a year to diminish by as much 20 – 45% a year. So the first insight is with regard to the *magnitude* of the reduction of IC designs: Should the number of IC designs continue to diminish at these rates then the entire climate of the semiconductor foundry, integrated design manufacturers, and EDA industries will dramatically and irrevocably change.

The second insight from this analysis is that these trends are currently maintained by several independent variables and small perturbations in the values of these variables could reverse these trends. This will be discussed more in Section 7.

2.2   The Application Pull for Programmable Platforms

Today's growth markets for electronic applications – consumer electronics, wireless electronics, and hand-held computing – have a number of common characteristics: They are very high volume applications and they require cost-efficient solutions that supply high performance computing, energy efficiency, and field/software programmability. General-purpose processors are poorly suited to meet the requirements of energy efficiency and competitive cost. ASICs are unable to provide sufficient programmability. As a result a variety of programmable platforms are emerging to meet these application requirements.

Another related application trend is the consolidation of electronic suppliers. The number of suppliers of common electronic systems such as personal computers has shrunk from its peak. The remaining end-system suppliers wish to rely on as few component suppliers as necessary. The result is the production of fewer, highly-integrated and flexible integrated circuits. Programmable platforms are a natural solution here as well.

In addition to the arguments for an application pull for programmable platforms there is one additional unique pull for using an ASIP as a building block. Use of general-purpose processors or proprietary reconfigurable fabrics as building blocks for programmable platforms typically requires licensing fees and/or a commitment to fabricate the platforms at a particular semiconductor foundry. Application developers wishing to fully own their intellectual property and to have full freedom to choose their semiconductor foundry are likely to create their own ASIP.

3.   What is the Next Design Discontinuity?

3.1   System Level Design

Reviewing Figure 1.1 the next design discontinuity seems easy to predict. After *transistor*, *gate*, and *register-transfer level* the next natural level of abstraction is the *behavioral level* or *system level*. Working at the system level does enable higher-productivity design entry. Simulating at the system level can offer 10-100X speed-ups in simulation relative to the register-transfer level. These observations led to several start-up companies as well as significant investments by major EDA vendors. The start-up companies and their tools included the Comdisco Signal Processing Workbench, Redwood Design Automation, and CADIS's COSSAP. In addition to acquiring Redwood Design Automation and Comdisco's Signal Processing Workbench, Cadence made a significant investment in the Virtual Component Co-design (VCC) tool. In addition to acquiring CADIS and the COSSAP tool Synopsys made a significant investment in Behavioral Compiler™ and hedged its bets by investing in both SystemC™ and System Studio.

Looking broadly at the EDA industry the bet on these system-level tools has turned out to be a losing one. Why? While these design approaches offered significant productivity improvements in *design entry*, as well as significant *verification* improvement due to simulation speed-ups, the intended move to system level overlooked one key component of a design discontinuity: "A fully reliable and predictable *implementation* approach that offers "single-pass" success " As tool builders focused on raising the level of abstraction they failed to notice that the foundation at the register-transfer level was crumbling beneath them. In fact, even the foundation of the gate-level netlist, upon which register-transfer level design had been built, was eroding.

As process geometries shrank in prior process generations, the primary issue for tool developers was the exponential increase in the number of devices that their tools had to cope with. As process geometries shrank below 250 nanometers, tools also had to begin to try to incorporate deep submicron effects such as interconnect capacitance, coupling capacitance, and the other effects mentioned in Section 2.1 above. Even today EDA tool vendors have not been entirely successful at providing a reliable flow to implementation. Today it is unthinkable that the "sign-off" package of the 1980's and 1990's (consisting of a standard-cell library, a netlist of cells from the library, a set of simulation vectors, and a set of manufacture test vectors) would be sufficient to ensure the manufacture of an IC that meets specification. In short, precisely when the industry needed a solid foundation to build upon, it disappeared.

3.2 Design through Intellectual Property (IP) Block Assembly

Another contender for fuelling a design discontinuity is IP-block assembly. The methods for design of IP blocks are codified in Bricaud and Keating's *Re-use Methodology Manual* [124]. In this approach design is principally performed through the assembly of pre-designed and pre-verified IP blocks around a standard bus such as ARM's Advanced Microprocessor Bus Architecture (AMBA™) or IBM's CoreConnect™ bus. Because blocks are individually verified, *verification* is eased. Because *design entry* is performed at a higher level it may be significantly more productive. Because *hard IP* blocks are already physically designed their *implementation* is accelerated. Even *soft IP* blocks come with tool scripts that accelerate their *implementation*. Based on these characteristics, design by IP-assembly looks very promising indeed. In fact, design reuse in general, and IP assembly in particular, is what has enabled the design of most complex SoC designs today.

The major EDA vendors have not neglected the promise of design by IP assembly. Intellectual Property is a major product category of Mentor Graphics. Synopsys IP block offerings, known as DesignWare®, is also a significant product offering of that company. Cadence put a lot of effort and credibility behind the Virtual Socket-Interconnect Alliance (VSIA). There are also a number of independent IP block providers including ARM, Rambus, and Tensilica.

Unlike system-level design, design by IP assembly does not have a single fatal flaw. Instead it suffers from a number of weaknesses that compounded together make the approach inadequate to truly provide the productivity boost needed in a design discontinuity. While IP blocks do simplify the *design entry* process, the resulting whole is not greater than its parts. Assemble a collection of IP blocks and what do you have? An assembly of IP blocks. If you want to really know what you have it is likely you will have to dig into the register-transfer level descriptions. The truth of this is exemplified by the fact that designing with an IP block most often requires access to the RTL description of the block. One approach to addressing this weakness is to complement design by IP assembly with a higher-level design philosophy known as *platform-based design* [126]. Platform-based design aims to provide an abstraction on top of the assembly of IP blocks, but it typically cannot provide a single integral view of the design.

Pre-verified IP blocks do promise to significantly reduce verification time. Unfortunately, the promise is often reneged on. IP blocks are notoriously unreliable. Many designers of SoCs have lamented that in the final assembly of the chip the one block that failed was the "pre-verified" IP block.

System level verification of a chip designed by IP assembly can also be challenging. IP blocks may be modeled at a variety of levels. To gain full confidence in the correctness of a design it may be necessary to model and simulate the entire design at the RT level. When this is true prior productivity gains will soon be lost.

Finally, because they are designed as a fixed layout, it may be difficult to integrate the implementation of *hard IP* blocks. Hard IP blocks are also difficult to migrate to smaller process geometries. On the other hand *soft IP* blocks can suffer from the same problem of RTL design in deep submicron geometries: Their precise physical properties are unknown until after place-and-route. This means that single-pass implementation is practically impossible to realize.

4. Programmable Platforms: The Next Design Discontinuity

Tailoring the overview of Section 1.2 specifically to the current RTL design approach, we can identify the key characteristics that a design methodology and toolset must provide to give impetus to the next design discontinuity relative to RTL synthesis:

- A *design-entry* approach that offers 10X improvement in productivity relative to logic synthesis from an HDL.

- A *functional verification* approach that offers 10-100X speed-up over simulation of an HDL at the register-transfer level.

- An *implementation* approach that is predictable and reliable.

We claim that programmable platforms are the most attractive approach to realizing these objectives. As noted in Section 3.2 of this chapter, platforms built of IP blocks go some distance toward realizing these objectives, but to fully realize the requisite productivity gains platforms must be user programmable; hence, the term *programmable platforms*.

Programmable platforms consist of individual programmable elements, memory sub-systems, and on-chip communication networks. Furthermore, these programmable platforms must offer a single integral *programming model* to the application developer. This integral approach contrasts with the programming models of many contemporary SoC platforms that consist of a RISC-processor programming environment, a DSP-processor programming environment, and a Verilog®-based simulation environment all held together with a SystemC™ wrapper.

While programmability provides the necessary application flexibility, it has traditionally come at a significant cost in speed and power consumption. As mentioned above, at the point of design discontinuity designers are often willing to sacrifice factors of 10X in speed or power; however, successful methodologies must offer the designer the ability to achieve speed comparable to the prior methodology when necessary. In our view it is this latter requirement that means that programmable platforms must be more than assemblages of RISC and CISC processor architectures. Such processors have no flexibility to compete with ASIC performance when necessary.

5. Contemporary Evidence

Recall that one of the indications that a design discontinuity is occurring is that a team of only a few designers working in the new methodology can outperform a design team of thirty engineers working in the old methodology. In November 1999 the chief architect of a new high-performance network processor visited the newly formed MESCAL research team. The network multiprocessor had six micro-engines capable of supporting twenty-four individual threads. At the visit, the architect spoke of a recent design win for the device. A "major networking company" had been working on an ASIC for a routing application. No sooner would the design team achieve timing closure on the ASIC then marketing would return with a new customer requirement. Even modest changes to the Verilog® model of the ASIC would result in starting another long trek to achieve timing closure. After evolving market requirements had blown the schedule one time too many, a few engineers snuck out to hear a seminar on programming the network processor. Working on a software implementation of their application the engineers were able to quickly arrive at an acceptable implementation. The ASIC project was scrapped. The situation was so analogous to the early days of logic synthesis replacing schematic entry that it seemed certain to us that we would see a lot more design wins for programmable platforms in the future.

We contend that ASIPs will become the building blocks of future multiprocessor programmable platforms, but ASIPs have existed as individual components for a while now: The earliest examples were probably the first fixed-point digital signal processors (DSPs) for audio applications available since 1982. These processors had support for basic audio-rate signal processing algorithms (FFT, filters) through specialized functional units (multiply-accumulators), memories (multiple data banks), and control functions (zero-overhead loops). However, there was no immediate proliferation of processors specialized for other domains. The next domain to see the use of ASIPs was video processing in the early 1990s. This resulted in some established processors for this domain – notably from Trimedia and TI (the C6000 series). Until this point the need for programmability was driven by the requirement for flexibility and not by the difficulty of ASIC design. Probably graphics processing was the next application area to benefit from ASIPs through Nvidia's development of graphics processors that appeared in 1995.

The last five years have seen the emergence of two additional domains with a strong representation of both ASIPs and programmable platforms – network and communication processing. At last count there were over thirty distinct network processor designs [219]. What is interesting about this space is that the processors are further sub-specialized based on the specific networking tasks that they are targeting, i.e. ASIPs are being targeted to narrower sub-domains such as packet forwarding.

While in the network processing space all ASIPs are instruction-set architecture (ISA) based, in the communication processors space we have a greater diversity of implementation styles. These cover a wide range from programmable hardware, such as the dynamically reconfigurable Chameleon and Quicksilver processors, to the Morphics VLIW architecture. The low end of this space has absorbed the traditional DSPs.

What characterizes this new breed of ASIPs in the networking and communication spaces is that unlike their predecessors, these ASIPs were created not just to provide flexibility through programmability, but in a large part, also to provide an easier implementation alternative to ASICs for their respective application domains. We expect this trend to grow significantly into other domains (and sub-domains as evidenced by the networking and communication spaces) in the near future.

A recent programmable platform is the IBM Cell processor. While general-purpose processors have routinely encroached on the territory of ASIPs by bringing lower performance applications onto the general-purpose processor, the high performance IBM Cell processor is reversing that trend. This multi-processor-on-a-chip was initially targeted for videogames, but is so powerful and flexible that it is being considered as a replacement for general-purpose processors in personal computers.

6.   Makimoto's Wave

Another rationale for the rise of programmable devices comes from the phenomenon known as "Makimoto's Wave" illustrated in Figure 1.2.

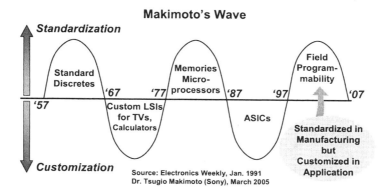

Figure 1.2.   Makimoto's Wave.

What Dr. Makimoto illustrates in his "wave-theory" is that there is a continuous battle between the drive for customization (e.g. ASICs) and standardization (e.g. ASIPs and programmable platforms).

The forces for standardization include:

1 New device structures – countering the need for product differentiation.

2 New architectures – countering the need for value added solutions.

 Software development innovation – addressing supply-demand imbalance.

The forces for customization include:

1 Improved design automation – addressing the need for productivity.

2 Improved computer-aided manufacturing – addressing the need for cost-effectiveness.

3 Improved computer aided testing – addressing the need for operational efficiency.

Makimoto's wave, first developed in the late 1980's, predicted a shift from ASICs to programmable devices in 1997. While the date seems to need adjustment the prediction of the trend is interesting. What may be even more interesting about Makimoto's wave is that it gives a simple formula for predicting a shift: When innovations in devices, architectures, and software development are greater than innovations in design automation, design for manufacturing, and technology-CAD, then standard solutions will predominate.

7.   Could We Be Wrong?

7.1   Could ASIPs be the Wrong Choice?

Because of Moore's Law, the semiconductor and EDA industries are in a constant state of chaos. We mean this in the very strict sense that examining a sufficiently small portion of the industry can support almost any trend. We would like to pause to consider some alternative perspectives:

1 The trend of diminishing numbers of ASIC designs is a temporary phenomenon:

  (a) As the semiconductor industry rebounds more revenue will be available for design and

  (b) Design approaches such as *structured ASIC* will dramatically reduce the cost of designing ASICs.

2 The number of ASICs will certainly diminish, and programmable platforms are a key element of future system solutions; however, ASIPs are not the building block. The market will ultimately build programmable platforms from:

  (a) Standard processors.

  (b) Finer-grained reconfigurable solutions such as FPGAs.

We will address these points successively:

Point 1: Based on Formulas 1 and 2, for the number of IC designs to increase either semiconductor revenues must increase at a significantly higher rate, or the cost of design must dramatically decrease. Adjusted for cyclical downturns, semiconductor revenues have indeed steadily increased. Part of the increase has come from the trend that semiconductor devices are providing a larger and larger portion of the value of the electronic systems in which they reside. In other words for each dollar spent on electronic systems, an increasing portion of that dollar is going to buy the semiconductor components in the system. To dramatically increase this portion it would be necessary to proportionately increase the revenues of the electronic systems industry; however, the high cumulative (7+%) growth rate of the electronic systems industry is already bumping up against the much slower growth rate of the world economy. In short, a sustained increase in the average (16%) growth rate of the semiconductor industry seems unlikely.

Another way to increase the number of IC designs is to diminish the design costs. A number of different technologies, such as *structured ASICs*, are emerging in an attempt to reduce the manufacturing costs of ICs; however, they appear to be focusing on the wrong variable in Formula 3. Reducing the manufacturing costs will not be sufficient to significantly reduce product development costs. Nor is it likely that structured ASIC designs will be able to meet the performance, energy, and cost requirements of many high-growth areas such as consumer markets.

If human-related design costs are the lion's share of the budget then these are the natural costs to try to reduce. Formula 4 breaks these down further. Since significantly higher productivity methodologies have not emerged, the total staff effort required for a project cannot be significantly reduced. To date the most effective approach for managing human-related design costs has been to move design projects to areas where the loaded cost per staff-year is less. Cost reductions of 75% (i.e. four engineers offshore for the price of one US engineer) have been reportedly realized. Rumor has it that in some areas of the world the EDA tool costs per engineer have been reduced to zero! We presume that this is a temporary phenomenon. A 75% reduction in human design costs is very significant, but it is a one-time improvement. Moreover, for a variety of reasons it seems likely that these human and EDA tool cost reductions will diminish over time. Some even argue that increasing design activity in places like India and China will eventually result in *increasing* labor costs.

Point 2a: It is certainly true that processor architectures are expensive to maintain and for some applications software development and porting is a much greater cost than processor development. These facts do indicate that it is likely that many enduring programmable platforms will be built from standard general purpose processors, such as those from the ARM® or PowerPC® families. Nevertheless, as the design examples in Part II of this book indicate, impressive improvements (2 – 30X) in energy-efficiency and performance can be achieved by tailoring processors to a specific application. While those application areas that have a significant software investment in legacy code are likely to choose multiprocessor platforms built up from their standard processors, embedded applications developed "from scratch" are more likely to explore alternatives using ASIPs as a building block. In fact, we have now seen repeatedly that embedded application developers are willing to migrate to new programmable devices if the devices provide adequate software support and considerable (5-10X) performance improvement over their current solutions. One general characteristic of methodologies that enable design discontinuities is that they offer some route to achieve the efficiencies of the prior methodology. ASIPs offer the designer the ability to realize the "bit-level" efficiency of the RTL or gate-level netlist when necessary. General-purpose solutions do not. The ideal is to have the computational efficiency of ASIPs with the same ease (or less) of writing software for general-purpose processors. Tools such as the XPRES™ compiler for the Tensilica Xtensa® ASIP described in Chapter 8 aim to provide exactly that.

Point 2b: FPGAs are programmable platforms that give high-level programmability via embedded processors and "soft" processors and achieve "bit-level" efficiency through their reconfigurable logic. As a result they seem to be likely candidates to be the basic fabric of programmable platforms of the future. In fact it seems certain that they will be an important and enduring platform for many applications.

There are two key weaknesses of FPGA based solutions. The first weakness is higher cost and higher power. FPGA-based solutions pay for their flexibility. Precise figures vary but it is easy to see FPGAs lose factors of 2-4X in speed and 10X (or greater) in area relative to ASSPs. The second weakness of FPGA-based solutions is the lack of an integral *programming model*. Mixing hardware (targeted toward the reconfigurable fabric) and software (targeted for embedded processors on the FPGA) solutions currently requires designing in multiple independent tool environments. To be specific, when programming the reconfigurable fabric you write Verilog® and "think like a hardware designer." When programming the on-chip embedded or soft processors you write C-code and "think like a software designer." Taken individually these are relatively low productivity environments; furthermore, integration of these environments together is problematic. Tools such as Xilinx's System Generator address productivity issues in writing Verilog®, but do not offer an approach for supporting software applications.

It seems certain that reconfigurable-logic solutions such as FPGAs will be enduring programmable platforms in their own right. It also seems certain that reconfigurable logic will be embedded onto other programmable platforms. To broaden the proliferation of FPGAs as a programmable platform will require improving performance and reducing cost. It will also require improving the tool set that sits on the desk of the FPGA designer.

7.2   When Will We Ever Know for Sure?

Real design discontinuities are impossible to miss. It now seems clear that complex programmable platforms are becoming the predominate method for delivering system functionality. What is less clear is whether they will be built up from general-purpose processors, reconfigurable fabrics, or ASIPs. If we concede that all three building blocks will survive then the question as to which predominates may seem academic. But there will be significant differences in the embedded software, IP, EDA, and semiconductor industries depending on which of these three solutions becomes the predominate building block.

The purpose of the discussion of this section, as well as that of Section 2, is to illustrate that relatively small (2-4X) differences in the productivity and final product efficiency will determine the predominate approach. In short, ASIPs are contenders to become the building blocks of the programmable platforms of the future but they are not certain winners. As stated, the ability for ASIP-based solutions to get design wins depends very much on their ability to provide an attractive point in the space of power-delay-cost-productivity trade-offs. If ASIP-based solutions fail to achieve attractive energy-delay design points they may lose design-wins to general-purpose processor solutions. If ASIP-based solutions are too difficult to program then they may lose their productivity advantage over FPGA-based solutions. Defining the key elements of a highly-productive design methodology that creates efficient ASIP designs is the purpose of this book. Our next step in identifying these key elements is to examine what are recurrent problems in the existing ASIP design methodologies.

In short, while ASIPs are strongly motivated by current design trends, to be successful requires execution excellence. This book is devoted to trying to capture the key elements of a methodology to create a successful ASIP that can then be used as an efficient building block for larger programmable platforms.

8.   Gaps in Existing Methodologies

Figure 1.3 shows twenty network processors based on their approaches towards parallelism. On this chart, we have also plotted iso-curves of designs that issue 8, 16, and 64 instructions per cycle. Given the relatively narrow range of applications for which these processors are targeted, the diversity is puzzling.

We conjecture that the diversity of architectures is not based on a diversity of applications but rather a diversity of backgrounds of the architects! This diagram, and the relatively short lives of many of these architectures, demonstrates that design of programmable platforms and ASIPs is still very much an art based on the design experience of a few experts, with a clear lack of a complete methodology. We see the following as being the key gaps in current design methodologies that need to be addressed:

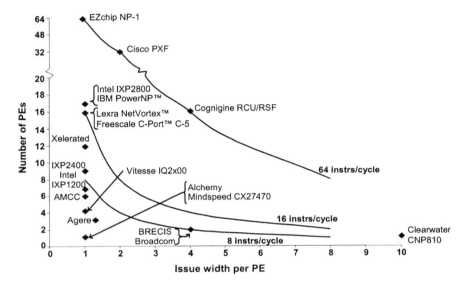

Figure 1.3. Diversity of network processors.

- *Incomplete application characterization:* Designs tend to be done with representative computation kernels, and without complete application characterization. Inevitably this leads to a mismatch between expected and delivered performance.

- *Ad-hoc design space definition and exploration:* We conjecture that each of the architectures represented in Figure 1.3 are the result of a relatively narrow definition of the design space to be explored, followed by an *ad hoc* exploration of the architectures in that space. In other words, the initial definition of the set of architectures to be explored is small and is typically based on the architect's prior experience. The subsequent exploration to determine which configuration among the small set of architectures is optimal is completely informal and performed with little or no tool support.

- *Inadequate software environments:* Simulators are typically manually generated *after* the architecture has been determined. Architectures are evaluated using assemblages of small code fragments. Higher-level language support, if it exists at all, is often an afterthought.

9. A Methodology for ASIP Design

The MESCAL research project is focused on developing a coherent methodology and set of supporting tools for the development and deployment of programmable platforms. Many programmable platform and ASIP development projects have avoided some or all of the pitfalls mentioned above; Our aim is to codify the best practices of ASIP design teams. What we have arrived at is a simple five-element methodology for approaching the design of ASIPs and programmable platforms. While this methodology is quite simple, we feel that an organized application of the methodology would improve many ASIP and programmable platform designs. While the approach is equally applicable to entire programmable platforms, in the remainder of the book we will focus on the development of individual ASIPs.

We call the factors of the methodology *elements* and not *steps* because the elements are not sequential. In fact an ASIP design is almost certain to be highly concurrent and iterative. For example an application choice may not only determine a set of benchmarks but also the required software environment for deployment. The support for a particular programming language in the software environment may strongly influence the architectural design space to be explored. Nevertheless, in the following book-style presentation we necessarily present the process linearly.

9.1 Element 1: Judiciously Using Benchmarking

ASIP development by definition must be application driven. Often the selection of the benchmarks driving the design receives inadequate attention. In general, benchmarks must be representative of the targeted application domain, and easy to specify and manage. They need to enable a quantitative comparison of the architectures being considered. Traditionally benchmarks have consisted of application kernels, and the only quantitative measure used is the number of execution cycles. This is inadequate.

A benchmark must consist of a functional specification, a requirements specification, an environmental specification, as well as a set of measurable performance metrics. The requirements specification recognizes that the application needs to meet certain requirements, and over-performing relative to these requirements does not add to design quality. The environmental specification ensures that any stimuli needed to exercise the benchmark are well specified. Finally the set of measurable metrics recognizes the need for additional metrics,

such as power consumption, memory utilization, and communication bandwidth requirements.

In the MESCAL methodology, benchmarks should be:

- Selected with careful application domain analysis.

- Indicative of real-world performance.

- Representative of both the environment and the functionality.

- Communicated with a precise specification.

In order to be indicative of real world performance, we find it essential that
the benchmark be at the application and not the kernel level. While kernel level
benchmarks are simple, they can potentially hide performance bottlenecks.

9.2   Element 2: Inclusively Identify the Architectural Space

The design space for ASIPs consists of a wide range of available choices
along multiple axes. A careful definition of the appropriate design space is
essential for subsequent design space exploration. While not all of this space
may be appropriate for a particular application domain, the initial description
of the design space should be *inclusive*. It seems unlikely that each of the
architects of the processors in Figure 1.3 considered the broad range of architectural choices available to him or her. We speculate that architects typically
restrict themselves to a small range of architectures with which they are already
familiar. If the design space is arbitrarily restricted from the beginning, then
many promising architectures may never be considered. For these reasons an
inclusive definition of the architectural design space is of utmost importance.

9.3   Element 3: Efficiently Describe and Evaluate the ASIPs

In order for a designer to consider a wide range of architectural alternatives,
architectures must be easy to describe and evaluate. If a new simulator and
software environment must be created for each point in the design space then
only a few points in the design space can possibly be evaluated. Systematic
exploration of the design space requires software tools to map the application
benchmark(s) onto the architectural design point in consideration, as well as
to evaluate the quality of this mapping for the design metrics. This requires
a software mapping and evaluation environment for each design point to be
considered. What is needed is a retargetable software environment, driven by
a common description of the design point in consideration.

9.4 Element 4: Comprehensively Explore the Design Space

Design space exploration is a feedback driven process where each design point is evaluated on the appropriate metrics. This is done using a software environment consisting of a compiler to map the application to the architecture, and simulators to quantify the quality of the result. Multi-objective searches can also be driven by synthesis using the generated HDL description. The key to doing this efficiently for a large number of design points is to quickly reconfigure the framework and simulators for each design point. The most direct way to accomplish this is to automatically generate retargetable compiler and simulators that start from a central description of the design point. Augmenting simulation-based design space exploration with exploration using analytical models can quickly prune the space of feasible solutions. The remaining area of the design space can be explored nearly exhaustively.

9.5 Element 5: Successfully Deploy the ASIP

An ASIP may achieve high efficiency by providing for an excellent match between the application and the architecture and at the same time may be completely useless if it cannot be programmed efficiently. Programming at the bit or assembly levels is unacceptable at current levels of software complexity. There is at least one instance of an ASIP project being canceled after successful silicon was available to customers, because of the sheer inability to program the chip efficiently. Efficient programming does not mean programming in C as is widely construed. It is hard, if not impossible, for existing and emerging compiler technology to fully harness the hardware capabilities of ASIPs starting from application descriptions in C.

Nonetheless, there are valuable lessons to be learned from the C language. C provided an efficient programming model for von Neumann architectures of the 70's and their modern derivatives. It accomplished this by exposing 20% of the architectural features of these processors (such as pointers and registers) that resulted in 80% of their performance. What we need is the modern equivalent of such a programming model that is capable of exporting critical features of these ASIPs that will enable exploitation of their specific features. How do we do this while retaining application portability? Do we need separate language features for different architecture classes?

10. What is Missing in This Book

10.1 A Comprehensive Survey of Related Work

While related work is surveyed in the individual chapters, as a whole this book does not attempt to give a comprehensive survey of past research on ASIP design techniques. Neither does it attempt to give a comprehensive survey of the burgeoning number of contemporary commercial and academic approaches to ASIP design. A comprehensive survey of related work would have doubled the size of the book; fortunately, each of these topics will be addressed in an upcoming book by Leupers and Ienne entitled *Application Specific Embedded Processors – Design Technologies and Applications*. We also presume a basic familiarity with the principles of architecture as described in [96] as well as basic principles of embedded systems [73] and the design of systems on a chip [246]. The focus of our book is simply to provide a methodology for contemporary ASIP designers and to illustrate that methodology with mature commercial tools that are currently available.

10.2   The Impact of Silicon Issues on ASIP Architecture Selection

While we could argue that elements two and three of our methodology: "Inclusively Identify the Architectural Space" and "Efficiently Describe and Evaluate the ASIPs" entail a thorough understanding of silicon implementation issues, we do feel that these issues have not received adequate attention in this book. Techniques for achieving high performance are addressed in [50] among others. The point that requires further clarification is the interaction between higher-level architectural concepts such as throughput and latency and lower-level implementation issues such as clock-cycle optimization. We hope to remedy this deficiency in the future.

10.3   Teams – The Human Element

ASIPs are not designed by tools or methodologies; ASIPs are designed by humans working in teams *using* tools and methodologies. The development of an ASIP will typically bring together individuals with diverse backgrounds: Application developers, software developers, computer architects, EDA tool experts, and integrated circuit designers. Successful development of an ASIP requires effective team management of this diverse group of individuals. This book does not address best practices for team management of ASIP projects; however, using the MESCAL methodology does help team management in two ways: First, lack of clarity surrounding roles and responsibilities is the source of many management problems. Following a well-defined methodology and assigning human ownership of elements of the methodology should help to clarify roles and responsibilities. Second, lack of consistency of representation is the source of many technical misunderstandings and errors in the design process. Each of the approaches described in this book generate multiple design views from a *single underlying representation* (see Figure 4.2 and Figure 7.5). In this way the architects, compiler writers, and the hardware implementation team all share a common representation of the design. Team management issues are explored further by Ron Wilson in "Non-Technical Issues in SoC Design" in [157].

10.4   From ASIPs to Programmable Platforms

The original scope of this book encompassed the design of programmable platforms and focused on the design of ASIPs as the basic building blocks of programmable platforms. As work continued two things became clear: First, covering the design of ASIPs alone would easily produce a book's worth of material. This is evidenced by the size of the work we have produced. Second, we felt that our (the MESCAL team's) insights into the development of communication issues in programmable platforms were not as developed as our insights into building ASIPs. There are numerous issues associated with building *networks-on-a-chip* (NoC) and multi-processor systems-on-a-chip (MPSoC) that we continue to explore and try to understand. While our collaborators at CoWare and Tensilica were ready to contribute their insights and approaches to building multi-processor programmable platforms, we opted to wait until we had more pages to explore the subject and a more mature perspective with which to fill the pages. In short, we hope to complement this book with a volume applying the MESCAL methodology to programmable platform design in the future.

11.   Summary

This book makes a case as to why programmable platforms will enable the next design discontinuity. In particular we argue that system designers looking to realize system applications will increasingly eschew ASICs designed through an HDL-based synthesis methodology for programmable platforms. These programmable platforms are themselves assemblages of programmable components. Chris Rowen, Tensilica's founder and CEO, has been quoted as saying: "The processor is the NAND gate of the future." We paraphrase this to say: "ASIPs are the standard-cells of the future." Both the gate-level design era of the 1980's and the register-transfer level design era of the 1990's relied heavily on well-designed standard cells. Similarly, we anticipate that the next design discontinuity will rely heavily on well-designed ASIPs as basic building blocks. However, we believe that the future success of ASIPs will depend on developing high-productivity design methodologies for ASIPs that produce efficient designs. Describing such a methodology is the goal of this book.

12.   Overview of the Remainder of the Book

While we ultimately aim to describe a methodology for building entire programmable platforms, in this book we focus on building ASIPs. In particular we outline a five-element methodology for ASIP development. In Part I this methodology is demonstrated with the Tipi approach. Tipi is a part of the MESCAL research project that is exploring a minimalist approach ASIP design. We mean minimalist both in the sense of a minimal toolset and in the sense that the individual ASIPs are very primitive processing elements. Tipi will demonstrate how these five elements provide a disciplined way for developing ASIPs.

In Part II we present contributed chapters that demonstrate the MESCAL methodology applied using advanced commercial tool sets for ASIP development from CoWare and Tensilica. These two companies produce tool sets that very much embody the state-of-the-art in tools for ASIP design, and their contribution of these chapters significantly enhances the utility of this book.

Chapter 7, contributed by Heinrich Meyr *et al.* of CoWare, presents the Embedded Processor Designer (LISATek[5]). This tool allows the user to jointly design the compiler and architecture of an ASIP. In LISATek[TM] the ASIP is captured by a *single* model, described in the architectural design language LISA, from which the tools (simulator, assembler, linker, compiler, RTL generator, and so on) are generated automatically. Once the architecture exploration phase is completed the same model generates the RTL description of the processor. LISATek[TM] allows the designer to explore the full architectural design space. The possibilities range from designing a simple computational element with only a few instructions to using advanced architectural concepts such as VLIW, SIMD, etc. in a complex processor.

The interconnect of the processing elements (ASIP) of a platform is as important as the elements themselves. A tool suite must be capable to model these networks-on-chip at various abstraction levels, both temporal and spatial, and it must seamlessly work together with the processor design tool. In the CoWare tool suite (ConvergenSC[TM]), analogously as in telecommunication networks, the communication services are are offered independently of the actual network topology. This allows the designer of a platform to explore multiprocessor alternatives.

In this chapter the LISATek[TM] approach is first generally overviewed and the links to the MESCAL methodology are given. In this presentation elements three, four, and five of the MESCAL methodology are treated as a single unit. In other words, the phases of efficiently implementing and analyzing the architecture, exploring the design space, and deploying the ASIP are presented as three very closely intertwined tasks.

---

[5]The LISATek[TM] tool suite is based on technology developed at RWTH Aachen University. It was commercialized by LISATek Inc. which has been acquired by CoWare in 2003.

The chapter concludes with ASIP-design case studies that illustrate various facets of the LISATek™ approach. The design of an ICORE architecture focuses on deriving optimized application-specific instructions. The ASMD and LISA 68HC11 architectures demonstrate the modeling and implementation efficiency of LISATek™. Finally, the ST200 case study covers the modeling of a high-end VLIW architecture.

Chapter 8, contributed by Grant Martin and Steve Leibson of Tensilica, first reviews their approach to creating ASIPs. Their approach is based on two configurable processor architectures. The instruction set of these architectures can then be extended by defining and compiling new instructions. An automated design flow uses the processor configuration and instruction extension definition to quickly and automatically produce both the hardware description and software tools. The hardware description consists of an HDL model together with scripts to guide the detailed implementation. The software tools include an instruction-set simulator, an assembler, and compiler for the new architecture. Using this high-degree of automation the designer is able to experiment over a wide space of ASIP architectural and configuration possibilities. The chapter then illustrates how the Tensilica approach is an exemplar for the MESCAL five-element methodology, with particular attention paid to benchmarking and design space exploration. In particular they show both manual and automated approaches to exploring the design space. While Tensilica's processor has numerous commercial design wins, to illustrate the breadth of their approach, they chose to demonstrate the performance of their configurable processors across a wide range of EEMBC [61] benchmarks.

The deployment section of the Tensilica chapter also goes beyond the single ASIP, ASICs and ASSPs and elaborates a number of possible multi-processor architectural approaches using Tensilica ASIPs as key design elements. In this way this chapter illustrates how entire programmable platforms can be built using ASIPs as a building block.

I

PART I: THE MESCAL METHODOLOGY FOR BUILDING ASIPS

Chapter 2

JUDICIOUSLY USING BENCHMARKING

Mel Tsai[1], Niraj Shah[1], Chidamber Kulkarni[1], Christian Sauer[2],
Matthias Gries[1], Kurt Keutzer[1]

[1]*University of California at Berkeley*
Electronics Research Laboratory

[2]*Infineon Technologies*
Corporate Research
Munich, Germany

Related work in benchmarking research usually focuses on uni-processor performance only. In this case, the optimization of computing performance is the primary concern. Common benchmarks contain an executable specification of the application with corresponding data sets for this purpose. We would like to extend benchmarking to evaluating full systems, such as programmable platforms. Due to the diversity and heterogeneity of system architectures and constantly evolving applications, it is currently a challenge to compare or evaluate systems quantitatively. In this chapter, we present principles of a benchmarking methodology that aim to facilitate the realistic evaluation and comparison of system architectures. A key aspect of our approach is the separation of function, environment, and the means of measurement within the benchmark specification. We pay particular attention to system-level interfaces and define a normalizing environment to quantitatively compare programmable platforms. Looking at the example of network processor architectures, we present a set of representative benchmarks and motivate their selection based on a taxonomy of network equipment. Finally, we provide a proof-of-concept of our methodology by specifying and implementing three benchmarks on a network processor.

1.   Introduction

The most established application domain of benchmarking is general-purpose and high-performance computing. In these domains benchmarks are mainly being used to evaluate the quality of computing resources. Application-specific platforms however, such as the Philips Nexperia™ for media processing and the Intel IXP for network processing, have to consider additional constraints on timing, costs, and power dissipation since they are usually used as embedded system. As a consequence of these tight constraints, their system architecture consists of several and heterogeneous components, varying from hard-wired co-processors to programmable cores. We can no longer assume that apart from computation all other components are overprovisioned, as general-purpose computing benchmarks do.

We believe that a disciplined methodology is needed to bring architectural aspects and constraints of the environment into benchmarking. Since programmable platforms can be customized to the anticipated application domain, and the requirements can vary considerably from application scenario to application scenario, the accuracy of the benchmark specification becomes increasingly important. Particularly, two important goals of system-level benchmarks are:

- They must provide results that are *comparable* across different platforms.

- They must provide results that are *representative* and *indicative* of real-world application performance.

Benchmarking plays a critical role in the architectural development process of application-specific instruction processors (ASIPs). In the development of an ASIP, architects can apply a benchmarking methodology that facilitates the evaluation of a prospective architecture and thus enables efficient exploration of the space of potential architectures.

Compared with traditional benchmarking of general purpose processors, we particularly recognize the following issues with providing a system-level benchmark for an application-specific domain:

- *Heterogeneous system resources* – Application-specific platforms achieve performance through a number of architectural mechanisms, which are not necessarily similar across different platforms. In order to maintain comparability of results, a benchmark specification cannot always allow complete freedom in the use of resources.

- *Performance classes* – In the realm of DSPs, for instance, one would not compare an inexpensive low-power DSP to the latest sub-gigahertz DSP designed for multi-channel base stations. These two DSPs are designed for different target systems and exist at two different power and price points. Yet, both processors can perform the same kind of algorithms.

- *Interfaces* – In the domain of network processing, for instance, networks utilize a number of transport media and protocols, including SONET, Frame Relay, ATM, Ethernet, FDDI, and Token Ring. Thus, network processors have become increasingly flexible in the types of network interfaces they support. Unfortunately, every NPU is different, and so are their supported interfaces.

- *Choosing the right metrics* – Platform performance can be measured in a number of ways. Typical metrics include throughput, latency, resource usage, and power consumption. However, there are a variety of other performance metrics that are useful. These include derived metrics, such as cost effectiveness (performance/cost), ease of programming and debugging, and API support. Thus, the set of possible ways to evaluate performance quickly becomes large, and the "right" evaluation criteria may differ across architects, system designers, and customers.

To meet the goals of system-level benchmarking, we present a methodology that allows benchmark results to be comparable, representative, and indicative of real-world application performance. Comparability of results is achieved by precise specification and separation of benchmark functionality, environment, and means of measurement. Functionality captures the important aspects of the benchmark's algorithmic core. In contrast, the environment supplies the normalizing test-bench in which the functionality resides and allows results to be compared across platforms. Finally, guidelines for the measurement of performance ensure that quantitative results are consistently measured across implementations.

The specification of benchmark functionality, environment, and means of measurement is only one part of the methodology. For benchmarks to be representative of real-world applications, a correct way to select benchmarks is important. Thus, not only does our methodology describe how to select a realistic suite of benchmarks based on an application-domain analysis, but also how to choose the appropriate granularity of a benchmark so that results will be indicative of real-world performance.

The rest of this chapter is organized as follows: Next, in Section 2 we describe our generalized benchmarking methodology. As a demonstration, we apply it to the network processing domain in Sections 3 and 4. Section 5 presents our proposed NPU benchmarking suite and includes an example benchmark specification for IPv4 packet forwarding. We demonstrate results for the Intel IXP1200 network processor in Section 5.3. In Section 6 we summarize previous and ongoing work relating to benchmarking. Finally, this chapter is concluded in Section 7.

2.   A Benchmarking Methodology

In this section, we motivate and present principles of a generalized benchmarking methodology that apply to programmable platforms in general. Next, we demonstrate these principles for network processors and illustrate our methodology on an IPv4 packet forwarding benchmark.

Before introducing our methodology, we first define the meaning of a benchmark.

DEFINITION 2.1 (BENCHMARK) *A benchmark is a mechanism that allows quantitative performance conclusions regarding a computing system to be drawn.*

Complete benchmarking methodologies produce benchmark results that are comparable, representative, indicative, and precisely specified, as discussed in the following sections.

2.1   Comparable

A quantitative performance claim that a benchmark provides is not enough. Benchmarks also need to be comparable across system implementations with heterogeneous interfaces. The benchmark specification cannot simply be a functional representation of the application; it must also model the system environment.

(1) *Quantitative benchmark results must be comparable across a range of system implementations.*

2.2   Representative

To properly characterize a particular application domain, a benchmark suite must cover most of the domain and provide results that correlate to real-world performance. While the number of benchmarks chosen should adequately represent the application domain, the usefulness and feasibility of implementation is questionable for large benchmark suites. A suite of benchmarks must be chosen based on careful application-domain analysis.

(2) *A set of benchmarks must be representative of the application domain.*

2.3   Indicative

A representative benchmark will not necessarily produce results that are indicative of real-world application performance. For benchmark results to be indicative, the benchmark must be specified with the right granularity. The granularity of a benchmark is its relative size and complexity compared to other applications in a particular domain. In the domain of network processing,

NetBench and NPF-BWG (see Section 6), for example, implement benchmarks according to three different granularities: Small (micro-level), medium (IP or function-level), and large (application or system-level).

The right benchmark granularity is determined by finding the performance bottlenecks of the application and architecture. In some cases, a particular subset of the application (e.g. a small application kernel) may dominate the performance of the larger application. However, in many cases bottlenecks cannot easily be identified or are heavily influenced by the architecture on which the application is implemented. In such cases, choosing a benchmark granularity that is too small may lead to performance results that are not indicative of real-world application performance.

(3) *The granularity of a benchmark must properly expose application and architectural bottlenecks of the domain.*

2.4   Precise Specification Method

A key component of our methodology is the precise specification of a benchmark. A benchmark specification requires functional, environment, and measurement specifications communicated in both a descriptive (e.g. English) and an executable description.

To separate concerns of benchmarking, we distinguish between functional, environment, and measurement specifications.

- The *functional specification* describes the algorithmic details and functional parameters required for the benchmark. This specification is independent of the architecture.

- In contrast, the *environment specification* describes the system-level interface for a specific architectural platform. This ensures comparability of results across multiple architectural platforms. The environment specification also defines the test-bench functionality, which includes stimuli generation.

- The *measurement specification* defines applicable performance metrics and the means to measure them. At a minimum, a method to determine functional correctness should be provided.

These specifications should be communicated in two forms: A descriptive (e.g. English) specification and an executable description. The English description should provide all of the necessary information to implement a benchmark on a particular system. This description also provides implementation guidelines, such as acceptable memory/speed trade-offs (e.g. it may be unreasonable for a benchmark to consume all on-chip memory), required precision of results, and acceptable use of special hardware acceleration units. Besides the

English description, the specification includes an executable description. This allows rapid initial evaluation, precise functional validation, and facilitates unambiguous communication of the requirements. Note that neither description eliminates the need for the other. We state the final tenet of our benchmarking methodology:

(4) *Each benchmark should include functional, environment, and measurement specifications in an English and executable description.*

3.   A Benchmarking Methodology for NPUs

Due to the heterogeneity of platform architectures, it is necessary to separate benchmarking concerns to ensure comparability of results. In this section, we present our benchmarking methodology for network processors as an illustrative example that separates the functional, environment, and measurement specifications.

3.1   Functional Specification

The role of the functional specification for our NPU benchmarking methodology is three-fold:

First, it must specify the benchmark requirements and constraints, such as the minimum allowable routing table size supported by the implementation, the minimum quality of service (QoS) that must be maintained, or restrictions on where tables can be stored. Requirements and constraints may vary significantly across the suite of benchmarks.

Second, the functional specification must describe the core algorithmic behavior of the benchmark along with any simplifying assumptions in an English description.

Finally, the functional specification should include an executable description written in the Click Modular Router Language [130]. While the algorithmic and functional behavior of a benchmark can unambiguously be communicated using standard languages such as C or C++, Click is a parallel, modular, and natural language for the rapid prototyping of network applications. Using Click, the programmer connects elements (application building blocks such as IP routing table lookups, queues, and packet sources/sinks) and runs simulations on this description to design and verify network applications. Click is also extensible; new elements can be added to serve the needs of different applications.

3.2   Environment Specification

Previous approaches to NPU benchmarking lack environment specifications that are critical for comparability. We present different models to specify the environment: The NPU line card model, the gateway model, and the NIC model.

These models overcome many of the difficulties associated with comparing benchmark results from different network processors.

Currently, major targets of network processors are core routers and edge network equipment. In these segments, routing and switching are the most common tasks. These tasks are performed by systems based on line cards with switched back-plane architectures. A close examination of how a network processor sits in a line card assists in defining the boundary between functionality and environment. Later we illustrate the gateway model, more suited to the low-end access segment of network equipment.

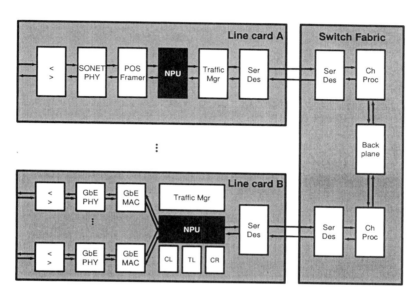

Figure 2.1. An example of a router architecture.

Line cards manage the physical network interfaces of a router. Typical router architectures, such as the one presented in Figure 2.1, contain a variable number of router line cards [52, 119]; up to 16 or more line cards are included in larger routers. Each line card is also connected to a switch fabric that passes packets to other line cards or control processors.

Figure 2.1 demonstrates two different line card deployment scenarios. The network processor in line card A is configured in a serial fashion to its surrounding components, while the network processor in line card B is connected to multiple Gigabit Ethernet MAC units in a parallel fashion. Network processors within line cards may have different interface configurations. In this example, line card A supports an OC-192 packet-over-SONET interface, while line card B supports 10x1 Gigabit Ethernet interfaces.

Table 2.1. Common router line card configurations.

Number of Ports	Type of Ports	Max Bandwidth (Aggregate), in Mb/s
16	Fast Ethernet	100 (1,600)
24	Fast Ethernet	100 (2,400)
48	Fast Ethernet	100 (4,800)
8	Gigabit Ethernet	1,000 (8,000)
16	Gigabit Ethernet	1,000 (16,000)
8	OC-3 POS	155 (1,242)
4	OC-12 POS	622 (2,488)
4	OC-12 ATM	622 (2,488)
1	OC-48 POS	2,488 (2,488)
4	OC-48 POS	2,488 (9,953)
1	OC-192	9,953 (9,953)

3.2.1   Interfaces.   In the environment specification it is vital to capture interface differences between line cards. Without capturing these aspects, comparability of benchmark results is not ensured because the network processor is not doing the same work if its line card configuration is different. In Figure 2.2, the system-level and architectural interfaces of an NPU are further illustrated. The system-level interfaces consist of the network (physical and fabric) and control interfaces; these interfaces directly influence the benchmark specification due to their tremendous impact on benchmark functionality and performance. Accordingly, this chapter focuses on system-level network interfaces of an NPU. The memory and coprocessor interfaces are architectural interfaces that are not visible at the system-level, and are not considered here.

Based on an examination of over twenty NPU interfaces [218], we found two major distinguishing characteristics: 1) Some NPUs integrate network interfaces, while others do not and 2) of those that do, the supported interfaces vary greatly. For instance, some network processors contain integrated MAC units for Ethernet interfaces and some do not. If the network processor does not include an integrated MAC unit, one must be added to the hardware or software simulator. If the network processor includes an integrated MAC, it must be configured to conform to the specification. The MAC unit(s) must support the required per-port data rate and number of ports. Further network interface restrictions may be specified by individual benchmarks (e.g. buffer size limits).

For the environment specification, the network interfaces in Figure 2.2 have three important parameters: The number of ports, the type of ports, and the speed of each port. The functional specification must also be made aware of these parameters as they directly impact functionality. The number and type of active ports in the system must specifically be defined in order to program where, when, and how packets flow within the network processor. In most cases, optimal performance cannot be achieved without customization of the benchmark software to suit a particular port configuration. For example, a programmer of an Intel IXP1200 must treat slow ports (i.e. 10/100 Fast Ethernet) significantly differently than fast ports (i.e. Gigabit Ethernet) [108].

3.2.2   Line Card Configurations.   Line card configurations from router vendors, such as Cisco [52] and Juniper Networks [119], have a wide range of performance requirements, as shown in Table 2.1. Hence, we propose two classes of line card port configurations - one that targets lower performance line card deployment scenarios and one that targets higher performance scenarios. For this reason, each benchmark should specify two valid port configurations: One that models a low-end line card configuration and one that models a high-end configuration.

While each benchmark may specify its own port configuration, a uniform configuration should ease implementation complexity. A standardized port configuration also facilitates the understanding of the overall performance of a network processor.

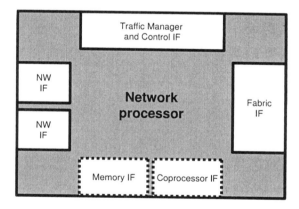

Figure 2.2.   Network processor interfaces.

For the low-end line card configuration, we recommend a standard configuration of 16 Fast Ethernet ports, a relatively low-bandwidth configuration supported by many router vendors. For the high-end line card configuration, we recommend a standard configuration of four OC-48 packet-over-SONET ports (10 Gb/s of aggregate bandwidth). OC-48 interfaces are common in high-end line card configurations [171].

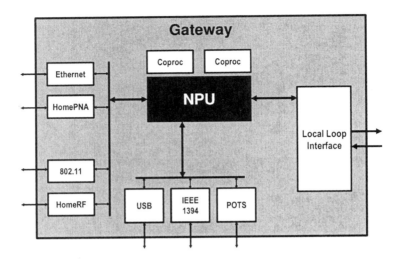

Figure 2.3.   Gateway environment model.

While the NPU line card model is a realistic environment for the core and edge segments, it does not accurately represent equipment within the access segment like gateways. A gateway acts as a boundary between an external network (e.g. the Internet) and an internal network (e.g. home or office network). Rich application services are performed in a gateway such as line termination, DHCP, network address translation, and firewall security. To represent this type of functionality and environment we define the gateway deployment scenario, presented in Figure 2.3. A handful of network processors target home and office gateway equipment [106, 34, 114].

The network processor inside the gateway acts as a centralized connection point for internal network devices to the local loop interface. The gateway connects multiple interface media such as wireless (e.g. 802.11b) and wireline (e.g. Ethernet) devices to an external broadband connection (e.g. xDSL or Cable).

Based on Figure 2.3, three important distinctions can be drawn between the gateway deployment scenario and the line card deployment scenario. First, in the gateway, the internal network interfaces are heterogeneous. Second, in addition to packet data, the gateway must also support non-packetized data from interfaces such as USB, IEEE 1394, or even analog phone line connections. Third, NPUs designed for gateway applications, such as those from Brecis [34], include payload-processing accelerators to support DSP (e.g. Voice over IP) and encryption (e.g. 3DES) tasks. A network processor without these accelerators or co-processors may be unable to perform such tasks. Thus, comparing network processors using the gateway model will not only require accounting for interface differences, but also differences in payload accelerators and co-processors.

In the future, a potential application of network processors is the Network Interface Card (NIC). A NIC connects a PC or workstation to a LAN, and has two primary interfaces: The network interface and the PC bus interface. Today, NICs are low-cost ASIC-based products designed for simple buffering and data transfer. However, evolving applications such as virtual private networks (VPN) at large data rates (>1Gb/s) will place greater processing requirements on the PC. Offloading of tasks to the NIC may be a necessary result of tomorrow's network processing requirements. An area of future work will be to develop the NIC environment model for network processors.

3.2.3   Traffic Sources and Sinks.   Additional considerations of the environment specification are the packet sources and sinks. The IETF Benchmarking Methodology Workgroup recommends [32] exercising network equipment using the following test setup: A tester supplies the device under test (DUT) with packets, which in turn sends its output back to the tester. In this case, the tester serves as both a packet source and sink. Alternately, a packet sender (source) supplies packets to the DUT, whose output is fed into a packet receiver (sink) for analysis. Either approach will serve the needs of our benchmarking methodology, as long as the packet traces meet the requirements of the benchmark environment specification.

According to the IETF, the data set for network interconnected devices, such as routers and switches, must include a range of packet sizes, from the minimum to maximum allowable packet size according to the particular application or network medium (i.e. Ethernet). Specifically, they recommend using evenly distributed Ethernet packet sizes of 64, 128, 256, 512, 1024, 1280, and 1518 bytes [32].

For protocols that use variable size packets, network processor vendors often report packet throughput using minimum-size packets. Because many network applications operate primarily on fields in the packet header, more processing must be performed on a large group of minimum-size packets than on a small group of large packets. Using only minimum-size or maximum-size packets can test the corner cases of benchmark performance, but these are not realistic indicators of overall performance and we defer to the IETF's recommendations and Newman's work for packet sizes.

Newman [171] observed packet sizes based on live Internet samples from the Merit network over a two-week period. The top four IP packet sizes were 40, 1500, 576, and 52 bytes (56%, 23%, 17%, and 5% of the total, respectively). Newman uses this proportion of packet sizes to develop a traffic pattern called the "Internet mix" (Imix).

Applications are still evolving in the gateway deployment scenario. Benchmarking for the gateway deployment scenario requires workload characterization and remains an open problem.

3.3   Measurement Specification

A benchmark implementation must be functionally correct before performance can be measured. In many cases, functional correctness can be observed by comparing the trace output of the network processor to the output of a reference implementation (e.g. the Click executable description). However, other measures of functional correctness (such as real-time constraints or allowable packet drop rate) may also be specified by the benchmark.

Once a benchmark implementation is shown to be functionally correct, its quality of results can be measured. For many network processor benchmarks, line speed (throughput) is the most important measurement of performance. There are two different units used to measure line speed: Packets per second and bits per second. While the former provides insight into computational performance, the latter provides a clearer understanding of throughput. However, with knowledge of the packet size distribution, one can be converted to the other.

According to RFC 1242 [31], throughput is defined as the maximum rate at which none of the offered packets are dropped by the device. We extend this definition so that some packets may be dropped in accordance with the benchmark specification. The measurement of throughput is challenging. According to [32], throughput is measured by sending packets to the DUT at a specific rate. If packets are incorrectly dropped by the DUT, the rate is successively throttled back until packets are no longer dropped. Using this procedure in our methodology ensures that the measurement of throughput is consistent across implementations.

Besides line speed, there are other useful performance metrics. For example, in a benchmark with real-time constraints, packet latency may be an important performance metric. Derived metrics such as cost effectiveness (i.e. performance/cost) may also be important. The notion of time is central to the measurement of many of these metrics. At the very least, a cycle-accurate software simulator of the network processor architecture is required.

First introduced by NPF-BWG [11] and discussed by Nemirovsky [170], the concept of headroom is loosely defined as the amount of available processing power that could be used by other tasks in addition to the core application. In theory, headroom is useful to evaluate support for additional functionality on top of the core application. Unfortunately, headroom is difficult to define and measure.

4.   An Example Benchmark Specification

To illustrate our methodology we present a benchmark specification in the domain of network processing for IPv4 packet forwarding. Using our template for network processor benchmarks shown in Figure 2.4, we motivate and summarize the specification for this benchmark.

4.1 Functional Specification

Our functional specification of this benchmark is based on RFC 1812, Requirements for IP Version 4 Routers [16]. The main points of the English description are:

Functional Specification

    Requirements and Constraints
    Behavior
    Click Implementation

Environment Specification

    Network Interface
    Control Interface
    Traffic Mix and Load Distribution

Measurement Specification

    Functional Correctness
    Quality of Results

Figure 2.4. Template for NPU benchmarking.

- A packet arriving on port P is to be examined and forwarded on a different port P'. The next-hop location that implies P' is determined through a longest prefix match (LPM) on the IPv4 destination address field. If P = P', the packet is flagged and forwarded to the control plane.

- Broadcast packets, packets with IP options, and packets with special IP sources and destinations are forwarded to the control plane.

- The packet header and payload are checked for validity, and packet header fields' checksum and TTL are updated.

- Packet queue sizes and buffers can be optimally configured for the network processor architecture unless large buffer sizes interfere with the ability to measure sustained performance.

- The network processor must maintain all non-fixed tables (i.e. tables for the LPM) in memory that can be updated with minimal intrusion to the application.

- Routing tables for the LPM should be able to address any valid IPv4 destination address and should support up next-hop information for up to 64,000 destinations simultaneously.

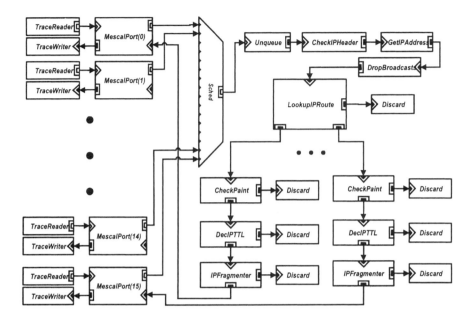

Figure 2.5. IPv4 packet forwarding Click diagram (low-end configuration).

In addition to the written description, Figure 2.5 shows the graphical representation of the accompanying Click description. This Click description also includes components of the environment specification (see below).

4.2   Environment Specification

For the network interfaces, we require the recommended 16 Fast Ethernet ports for the low-end configuration and four OC-48 POS ports for the high-end configuration. A simplifying assumption of our benchmark is that it is a stand-alone configuration. As a result, we do not require a fabric interface.

We define a simple control interface that drops all incoming control packets and does not generate any control packets.

The traffic mix for our benchmark contains destination addresses evenly distributed across the IPv4 32-bit address space. Packet sizes are evenly distributed across 64, 128, 256, 512, 1024, 1280, and 1518 bytes, and 1% of the packets are hardware broadcast packets [32]. In addition, 1% of the packets generate IP errors.

There is a single packet source for each input port that generates an evenly distributed load. Also, the range of destination addresses and associated next-hop destinations provide an evenly distributed load on every output port. Figure 2.5

contains packet sources and sinks that model network interfaces of the low-end configuration.

4.3   Performance Measurement

In order to prove the functional correctness of a benchmark implementation, one must compare all output packets (including error packets) from the Click description to that of the DUT.

Overall performance should be measured as the aggregate packet-forwarding throughput of the network processor such that no valid packets are dropped. This is measured by successively lowering the bandwidth at which packets are sent to each MAC port simultaneously until the network processor does not drop valid packets. This measurement is obtained by using packet traces and port loads specified by the environment.

5.   The NPU Benchmark Suite

After having shown the specification of one particular benchmark, we would like to define a representative set of benchmarks for a defined deployment scenario. Before we introduce our suite of benchmarks for network processors, we first discuss benchmark granularity for our methodology.

5.1   Granularity

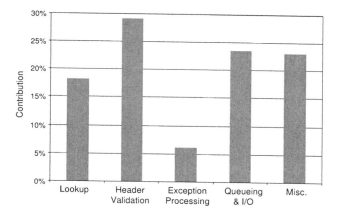

Figure 2.6. Profile for IPv4 packet forwarding on the IXP1200 (64 byte packets).

As introduced in Subsection 2.3, the appropriate benchmark granularity must be chosen after careful application-domain analysis and the identification of performance bottlenecks. Previous benchmarking approaches often chose benchmarks with relatively small granularities, e.g. micro-level. In our experience, typical applications running on a network processor must perform a large number of diverse tasks on packets. Nearly all such applications must perform micro-level operations, such as input queuing, table lookups, bit-field replacements, filtering rule application, appending bytes to packets (resizing), and checksum calculations. While it is possible that any one of these functions may become a performance bottleneck, it is difficult to determine a priori which micro-level tasks will dominate a particular application on a particular network processor. As an example, Figure 2.6 shows a micro-level breakdown of IPv4 packet forwarding benchmark on the IXP1200 for packet traces consisting only of 64 byte packets. Even when processing small packets, which heavily stresses the header processing ability of a network processor, LPM routing lookup consumes only 18% of the processing time on an Intel IXP1200 microengine. As Figure 2.6 shows, no single micro-level task dominates execution time, thereby supporting the chosen granularity of this benchmark.

We believe the performance bottlenecks of network applications are not appropriately represented by micro-level or function-level benchmarks. Hence, network processor benchmarks should be specified at the application-level.

5.2   Choosing the Benchmarks

Some existing approaches to NPU benchmarking define criteria for differentiating the characteristics of benchmarks (i.e. header vs. payload) and choose benchmarks based on these criteria. However, we strongly believe that the choice of benchmarks should be driven by an application-domain analysis. For this analysis, we created a classification of applications based on network equipment and chose benchmarks that were characteristic of these classes. This equipment-centric view provides a representative set of applications to evaluate system architectures.

We identify three major segments within the network application domain. First, equipment designed for the Internet *core segment* are generally the highest throughput devices found in networking. Core equipment is usually responsible for routing high-speed WAN/MAN traffic and handling Internet core routing protocols, such as BGP.

Second, the *network edge segment* comprises equipment that operates at lower speeds (with shorter links) than equipment in the core. These devices include a variety of mid-range MAN packet processors, such as routers, switches, bridges, traffic aggregators, layer 3-7 devices (such as web switches and load balancers), and VPNs/firewalls.

Finally, the *network access segment* is usually comprised of low- to high-speed LAN equipment. This segment includes equipment such as LAN switches (Fast Ethernet, Gigabit Ethernet, FDDI, Token Ring, and so on), wireless de-

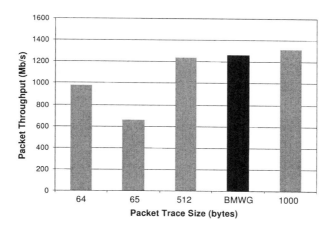

Figure 2.7. IPv4 packet forwarding throughput on the IXP1200.

vices (802.11 a & b), integrated access devices, access concentrators, and cable modems.

Based on this classification, we identify five major application categories in which network processors are likely to play a role. Ten benchmarks were chosen based on their relative significance within the network.

- *LAN/WAN Packet Switching*

    - Ethernet Bridge
    - IPv4 Packet Forwarding
    - ATM Switch
    - MPLS Label Edge Router

- *Layer 3+ Switching*

    - Network Address Port Translation (NAPT)
    - HTTP (Layer-7) Switch

- *Bridging and Aggregation*

    - IP over ATM
    - Packet over SONET Framer

- *QoS and Traffic Management*

    - IPv4 Packet Forwarding with QoS and Statistics Gathering

- *Firewalls and Security*

    - IPv4 Packet Forwarding with Tunneling and Encryption

5.3   Benchmarking Results for Intel IXP

We have completed three benchmark specifications for demonstrating the benchmarking suite: Apart from IPv4 packet forwarding described earlier, we also implemented MPLS and NAPT. These benchmarks were implemented in assembly language on the Intel IXP1200 network processor.

To specify our benchmark functionality and environment we augment an English language description with an executable specification in the Click Modular Router framework [130]. Describing network processor benchmarks using Click has certain key advantages. First, the intended functionality of the benchmark can be verified using Click's Linux®-based simulation tools. Second, the extensibility of Click allowed us to write a number of new elements and tools. For example, we have written a port element that models the ports of a router line card. Port elements are used in conjunction with new elements that read and write packet trace files (in a format similar to TCPDump [158]) and supply the Click simulator with realistic packet data from our packet-generation utility (written in C).

The same packet trace interface used by the Click simulator is used for the network processor simulation environment, which aids verification of benchmark functionality. In our benchmark experiments with the IXP1200, a Windows dynamic-linked library was written to interface with the IXP1200 Developers Workbench tools.

Performance on the IXP1200 was measured using version 2.0 of the Developer Workbench software assuming a microengine clock rate of 200 MHz and an IX Bus clock rate of 83 MHz. Intel's IXF440 MAC devices are modeled within the Developer Workbench and configured for 16 Fast Ethernet ports.

Figure 2.7 shows the results for IPv4 packet forwarding. A variety of packet sizes were tested, including the mix of packet sizes recommended by the IETF Benchmarking Methodology Workgroup (BMWG) in [32]. With 16 100 Mb/s ports, the maximum achievable throughput is 1600 Mb/s, yet for the range of packet sizes the IXP1200 cannot sustain this throughput due to the computational complexity of the benchmark specification.

As shown in Figure 2.8, the IXP1200 achieves noticeably higher packet throughput for the range of packet sizes on Network Address Port Translation (NAPT). Incoming packet headers are used to hash into an SRAM-based session table, an operation that benefits greatly from the IXP1200's on-chip hash engine.

As shown in Figure 2.9, the IXP1200 achieves the highest performance on our MPLS benchmark. This is due to the lower computational requirements of our MPLS specification compared to the IPv4 and NAPT benchmarks. This benchmark exercises all three modes of an MPLS router (ingress, egress, and transit) in a distributed fashion.

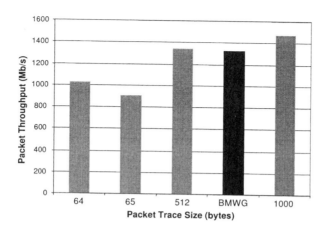

Figure 2.8.   NAPT on the IXP1200.

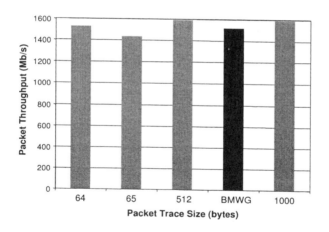

Figure 2.9.   MPLS on the IXP1200.

In theory, as packet size increases, packet throughput should also increase. In practice, we observe a reduction in throughput between 64 and 65 byte packets, 128 and 129 byte packets, and so on across all three benchmarks. This is due to the 64 byte alignment of the receive and transmit FIFOs on the IXP1200. Extra processing and transmission time are required for non-aligned packet segments.

This type of information provides insight into the architectural nuances of network processors. It can be seen that actually several different benchmarks representing the application domain are required to analyze the sensitivity of the platform to certain architectural choices. Judiciously using benchmarking is thus an integral part of a systematic development of ASIPs and programmable platforms.

6.   Related Work on Benchmarking

Available benchmarks can be classified according to their application domain. Examples are:

- *General purpose computing:* Benchmarks for general-purpose and scientific computing are published by the Standard Performance Evaluation Corp. (SPEC[1]). The Business Applications Performance Corporation (BAPCo[2]) focuses on benchmarks for personal computers and notebooks.

- *Embedded systems:* Benchmarks for embedded systems including automotive, telecommunication, consumer, and control systems can be found in MiBench [91] and are also defined by the Embedded Microprocessor Benchmark Consortium (EEMBC [61]).

- *Multimedia-centric computing:* Benchmarks focusing on multi-media processing can be found in MediaBench [142] and in the DSP-centric BDTI benchmarks from Berkeley Design Technology, Inc.

- *Database and transaction processing:* Business-oriented transactional server benchmarks are defined by the Transaction Processing Performance Council (TPC[3]). SPECjAppServer2002 is a client/server benchmark from SPEC for measuring the performance of Java enterprise application servers in end-to-end web applications.

- *Parallel Computing:* Examples of benchmarks for parallel computing and multi-processing are SPEC OMP (OpenMP Benchmark Suite), Stanford Parallel Applications for Shared Memory (SPLASH2 [247]), and PARallel Kernels and BENCHmarks (Parkbench[4]).

- *Network processing:* Related work on defining benchmarks for network processors include CommBench [245], NetBench [160], and activities of the Network Processor Forum (NPF[5]). Related work on disciplined approaches for evaluating network processors can be found in [45]. The Internet Engineering Task Force (IETF) has a work group on benchmarking methodologies (BMWG) of internetworking technologies.

---

[1] *http://www.spec.org*
[2] *http://www.bapco.com*
[3] *http://www.tpc.org*
[4] *http://www.netlib.org/parkbench*
[5] *http://www.npforum.org*

Popular benchmarking approaches, such as SPEC and MediaBench, work well because their intended architectural platform is homogeneous. In other words, these and other related approaches focus solely on the functional characteristics of the application and do not emphasize the system-level aspects (e.g. interfaces) of the architectural platform. As a result, these approaches do not work well for application domains with widely heterogeneous architectures.

Benchmarking efforts related to system-level architectures can be found in the context of embedded systems and network processing: CommBench, EEMBC, MiBench, NetBench, NPF Benchmarking Working Group, Linley-Bench, and by Intel Corp. In general, these approaches do not emphasize the system-level interfaces of a system. Since the performance of, for instance, a network processor is heavily dependent on system-level interfaces, such as the network and control interfaces, it is crucial to account for such differences in a corresponding benchmarking methodology. For example, how does one compare the Freescale C-Port^{TM} C-5 (which contains a programmable and flexible on-chip MAC unit) to the Intel IXP1200 (which does not contain an on-chip MAC unit)? Without special consideration of these environmental issues, fair comparisons between the C-Port^{TM} C-5 and the IXP1200 cannot be drawn. As determined earlier, judiciously using benchmarking also requires a specification methodology, an executable description, a method for performance measurement, and a motivation for selection of benchmarks. In order to underpin the necessity of the principles, we will have a closer look at some of the more recent benchmarking efforts.

CommBench

CommBench [245] specifies eight network processor benchmarks, four of which are classified as "header-processing applications" and the other four as "payload-processing applications". CommBench motivates their choice of benchmarks, but provides no methodology of specification or consideration of system-level interfaces. In addition, the implementation and analysis of CommBench currently targets general-purpose uniprocessors. It is unclear how their benchmark suite can be applied in the context of heterogeneous platforms.

EEMBC

The Embedded Microprocessor Benchmark Consortium (EEMBC [61]) defines a set of 34 application benchmarks in the areas of automotive/industrial, consumer, networking, office automation, and telecommunication domains. Due to this wide focus, the benchmark suits per application domain are not necessarily representative. In the networking domain, for instance, EEMBC defines only three simple benchmarks (Patricia route lookup, Dijkstra's OSPF algorithm, and packet flow between queues) with little notion of a methodology that can be applied to network processors. The benchmarks include a measurement specification.

MiBench

MiBench [91] follows a concept similar to EEMBC, but is composed from freely available sources. A set of 36 task-level benchmarks is defined in areas of automotive/industrial, consumer, office, networking, security, and telecommunication. The benchmarks include small and large data sets for each kernel as workload.

NetBench

NetBench [160] defines and classifies a set of nine network processor benchmarks according to micro-level (e.g. CRC32), IP-level (e.g. IPv4 routing), and application-level (e.g. encryption) granularities. By using this classification, NetBench acknowledges the heterogeneous aspects of network processor micro-architectures and has results for the SimpleScalar [13] simulator and three results for the Intel IXP1200. However, as with EEMBC and [245], NetBench does not provide a methodology that considers the heterogeneity of platform interfaces.

NPF Benchmarking Working Group

The NPF Benchmarking Working Group (NPF-BWG, [11]) defines benchmarks based on micro-level, function-level, and system-level applications. As of this writing, three function-level benchmarks are available: IP forwarding, MPLS, and a switch fabric benchmark. NPF-BWG addresses environment and system-related benchmarking issues, such as workload and measurement specifications. However, no source code is available.

LinleyBench

The Linley Group benchmark[6] builds on the IPv4 definition of the NPF-BWG. In addition, DiffServ classification and marking, and an optional ATM-IPv4 interworking test are defined. To date, results on the IBM network processor are published.

Intel Corporation

Intel's [45] benchmarking approach focuses on the IXP1200 network processor. They define a four-level benchmark hierarchy: Hardware-level, micro-

---

[6]*http://www.linleygroup.com/benchmark*

level, function-level, system-level. Hardware-level benchmarks are designed to test functionality specific to a particular NPU (e.g. memory latencies). They provide some results for IPv4 forwarding on the Intel IXP1200. Intel's executable description and method for performance measurement are specific to the IXP1200.

Intel's work recognizes the need for benchmark comparability by describing a benchmarking reference platform. This model wraps black-box functionality (the network processor, co-processors, and memory) with a set of media, fabric, and control interfaces. However, these interfaces are left unspecified so that customers can modify the reference platform to suit their particular needs. As we have argued earlier, these interfaces must be appropriately specified because they are critical for benchmark comparability. Intel's published work does not indicate whether they provide a methodology for benchmark specification.

NpBench

Lee and John [141] address the lack of control plane benchmarks for network processing. They define a set of 26 functions in three classes: Traffic management and Quality of Service, security and media processing, and packet processing). These benchmarks also include processing in the data plane and refer to CommBench, EEMBC, MiBench for this purpose.

Other Work

Nemirovsky [170] recognizes the needs and requirements of network processor benchmarking. Although Nemirovsky provides insight into the methods for quantifying network processor performance, he does not provide precise information about a viable benchmarking approach.

Crowley et al. [55] present an evaluation of theoretical network processor architectures based on a programmable network interface model and detailed workload analysis. However, their model does not include many aspects of current target systems for network processors. Also, their model does not fully separate functional and environment concerns.

Summary of Existing Approaches

In Table 2.2 we compare existing approaches to platform benchmarking, particularly for network processors. In this domain, the environment specification includes the specification of external network traffic. The granularity column distinguishes between micro-kernel benchmarks that focus only on the most compute-intense part of an application and more complex function benchmarks.

As this table shows, none of the existing approaches meet all the requirements of a benchmarking methodology. In particular, there are a number of deficien-

Table 2.2.   Characteristics of benchmark definitions.

	CommB.	NetBench	EEMBC	NPF	Intel	Linley	MiBench	NpBench
Measurement Specification	No	No	Yes	Yes	IXP1200-specific	Yes	No	No
Traffic Spec.	No	Yes	Yes	Yes	N.A.	Yes	Yes	no
Interface Spec.	No	No	No	N.A.	Yes	No	No	No
Source Code	C	C	C	No	μE-C	No	C	C
Availability	on req.	free	license	spec only	No	license	free	on request
Granularity	micro	micro/func	micro	function	function	function	micro	micro

cies including: Lack of specification methodology and lack of consideration of interfaces. In order to compare different platforms and achieve functional correctness, executable system benchmarks, environment, and measurement specifications are required.

7.   Conclusion

In this chapter we have reviewed existing benchmarking efforts and revealed flaws why they cannot be applied to benchmarking of whole systems. We have derived four principles of a generalized benchmarking methodology considering heterogeneous system architectures:

- Comparability across different system implementations,

- Representativeness of the application domain,

- Indicative of real performance,

- Precise specification.

The methodology defines a template for a benchmark consisting of functional, environment, and measurement specifications. As an illustrative example, we have tailored this methodology to the specific requirements and goals of network processor benchmarking. However, other application domains (e.g. telecommunications, multimedia, automotive) also benefit from this work.

Our key contributions, illustrated for network processing, are the emphasis of the system-level interfaces for NPUs based on our line card model, the utility of a Click executable description, motivation for application-level benchmarks, and a disciplined approach to identifying a set of benchmarks. In summary, we believe we have identified key issues in benchmarking ASIPs as well as programmable platforms and embodied them in a methodology for judiciously using benchmarking.

Chapter 3

INCLUSIVELY IDENTIFYING THE ARCHITECTURAL SPACE

Niraj Shah[1], Christian Sauer[2], Matthias Gries[1], Yujia Jin[1], Matthew Moskewicz[1], Kurt Keutzer[1]

[1] *University of California at Berkeley*
Electronics Research Laboratory

[2] *Infineon Technologies*
Corporate Research
Munich, Germany

The design space for a programmable platform is defined by a wide range of available choices along multiple axes (e.g. degree of parallel processing, special-purpose hardware, on-chip communication mechanisms, memory architectures and peripherals). However, the comprehensive description of all the axes is usually too complex for a subsequent exploration of the design space. Moreover, not all of this space may be appropriate for a particular application domain. Therefore, a tight definition of the appropriate design space is essential for finding an optimal solution in a feasible amount of time.

This chapter gives a comprehensive overview of the relevant design axes for programmable platform design. We introduce a discipline for inclusively identifying the essential design subspace to explore for a given application domain. The discipline can be based on a comprehensive analysis of existing designs and applications, or it may use benchmarks on hypothetical platforms to identify the sensitivity of the design to certain design decisions. We use network processing and memory subsystems as examples of relevant architectural spaces for programmable platforms.

We start with a survey of relevant design axes for programmable platform design in the following section. We continue in Section 2 with the exemplary survey of design axes for two subspaces: Memory design and peripherals. We do this to point out that the design space can be defined hierarchically, and that subspaces are complex and driven by different optimization metrics. In Section 3, we use the domain of network processing to illustrate how many design axes are relevant and how complex the design space can become for real-world designs. This example underpins the importance of identifying the design space systematically in order to make disciplined design space exploration feasible.

1.   The ASIP Landscape

The ASIP landscape provides us with a wide range of implementation alternatives. Each point in the space is an attempt to match application characteristics with hardware support to minimize the power and performance overhead of a programmable solution. At the top level we can broadly classify ASIPs into two classes:

- *Instruction Set Architecture (ISA) based:* These ASIPs have their roots in classic ISA processors. Efficiency enhancements are provided through specialized resources.

- *Programmable hardware based:* These ASIPs have their roots in configurable logic such as FPGAs. Mapping an application to such an ASIP is a hardware design task similar to programming an FPGA.

The above classification is not very clear for a couple of reasons. Increasingly, ASIPs are available with hybrids of the above forms of programmability (i.e. in the form of processors and programmable hardware on the same die). Further, as we will show, the distinction between an instruction and hardware programming bits is gradually getting blurred. We now examine each of these two main categories in terms of how they provide application specific support.

1.1   ISA Based ASIPs

Elements of this class can be distinguished from each other in terms of their approaches to parallel processing, elements of special-purpose hardware, structure of memory architectures, types of on-chip communication mechanisms and use of peripherals [219].

1.1.1   Parallel Processing.   Hardware is inherently parallel. However, a classic von Neumann processor is inherently sequential since instructions are executed one at a time. Thus, if ISA architectures are to approach the efficiency of ASICs, they must exploit all possible forms of parallelism. We see this happening at three different levels: The processing element (PE) level, instruction level and word/bit level. While these approaches are orthogonal, a decision at one level clearly affects the others.

Processing Element Level.   Multiple PEs (each executing its own instruction stream) can be used in the following two configurations:

- *Pipelined:* Each processor is designed for a particular processing task.

- *Symmetric:* Each PE is able to perform similar functionality.

In the pipelined approach, inter-PE communication is very similar to dataflow processing – once a PE is finished processing some data, it sends it to the next downstream element. Examples of this architectural style in the network processor domain include Cisco's PXF, EZChip's NP-1, Vitesse's IQ2000 and Xelerated Packet Devices. In general, these architectures are easier to program since the communication between programs on different PEs is restricted by the architecture.

Symmetric PEs are normally programmed to perform similar functionality. These are often combined with numerous co-processors to accelerate specific types of computation. Arbitration units are often required to control access to the many shared resources. Again, looking at the network processor space, the Cognigine, Intel IXP1200, IBM PowerNP^{TM} and Lexra NetVortex^{TM} are examples of this type of macro-architecture. While these architectures have more flexibility, they are difficult to use since programming them is very similar to the generic multi-processor programming problem.

The independent execution streams here may share the same memory space as in simultaneous multi-threading or be independent processes.

Instruction Level.   For multiple issue architectures, there are two main tactics for determining the available parallelism: at compile time (e.g. Very Long Instruction Word or VLIW) or at run time (e.g. superscalar). While superscalar architectures have had success in exploiting parallelism in general-purpose architectures (e.g. Pentium®), VLIW architectures have been effectively used in domains where compilers are able to extract enough parallelism (e.g. signal processing). VLIW architectures are often preferred because they are lower power architectures. In network processors, the Agere Routing Switch Processor, Brecis' MSP5000 and Cisco's PXF use VLIW architectures. In the communication processors space, the Improv Systems' Jazz has a customizable VLIW processor (typical issue width is eight). Clearwater Networks uses a multiple-issue superscalar architecture in the CNP810 series.

Bit Level.   Depending on the data types and operations present in an application, it is possible to exploit bit level parallelism. An example of this is an instruction that computes the CRC field of a packet header in network processing. In the absence of such an instruction, it would take a large number of instructions to sequentially go through each bit for this processing.

1.1.2   Special-Purpose Hardware.   A fairly direct way to approach the efficiency of ASICs is to have part of the processor resemble an ASIC by implementing common functions in dedicated hardware. This specialization comes at the cost of generality – and may not benefit all algorithms in the application domain. Further, the actual speedup obtained is limited by Amdahl's law (i.e, it depends on the fraction of total compute time that is being replaced by the hardware unit). Special-purpose hardware can be broadly divided into two categories: Co-processors and special function units.

Co-Processors.   A co-processor is a computational block that is triggered by a PE (i.e., it does not have an instruction decode unit) and computes results asynchronously. In general, a co-processor is used for more complicated tasks, may store state and may have direct access to memories and buses. As a result of its increased complexity, it is more likely to be shared among multiple processing elements. A co-processor may be accessed via a memory map, special instructions or a bus transaction. Operations ideally suited for co-processor implementation are well defined, expensive and/or cumbersome to execute within an instruction set. Furthermore, these operations are prohibitively expensive to implement as an independent special function unit. The functions of co-processors vary from algorithm-dependent operations to entire kernels. For example, the Hash Engine in the Intel IXP1200 is only useful for lookup if the algorithm employed requires hashing.

Special Function Units.   A special function unit is a computational block that computes a result within the pipeline stage of a processing element. The focus here is on operations for which it would take a large number of cycles for the computation with a standard instruction set. For example, Intel's IXP1200 has a single cycle instruction to find the first bit set in a register.

1.1.3   Memory Architectures.   A third strategy employed to improve efficiency is to change the structure of memory architectures. The major memory related tactics are: Multi-threading, memory management, and task-specific memories.

The most common approach to hiding memory latency is multi-threading. The stalls associated with memory access are well known to waste many valuable cycles. Multi-threading allows the hardware to be used for processing other streams while another thread waits for a memory access (or a co-processor or another thread). Many network processors (Agere, AMCC, ClearSpeed, Cognigine, Intel, Lexra, and Vitesse) have separate register banks for different threads and hardware units to schedule threads and swap them with zero overhead. Without dedicated hardware support, the cost of operating system multithreading would dominate compute time since the entire state of the machine would need to be stored and a new one loaded.

With memory management support, the goal is to eliminate the overhead of operating system calls by providing direct hardware support. This can be done using copy units for copying blocks of memory (e.g. Clearwater's Packet Management Unit), or as hardware list data structures for keeping track of memory blocks (e.g. Intel IXP 1200).

Like co-processors and special function units, which are specializations of a generic computational element for a specific purpose, task-specific memories are blocks of memory coupled with some logic for specific storage applications. For example, Xelerated Packet Devices has an internal CAM (Content Addressable Memory) for classification. On the Vitesse IQ2200, the Smart Buffer Module manages packets from the time they are processed until they are sent to an output port.

1.1.4   On-Chip Communication Mechanisms.   As the number of components in ASIP architectures grows, the type of on-chip connection and communication schemes for processing elements, co-processors, memories and peripherals becomes increasingly important. The major parameters of on-chip communication schemes are:

- Topology - flat or hierarchical.

- Connectivity - point-to-point or shared.

- Link type - bit width, throughput, and sharing mechanisms.

- Programmability - hardwired, configurable or programmable.

1.1.5   Support for Peripherals.   In addition to support for computation, ASIPs may also provide support for getting data in and out of the processing core. This peripheral support may come either in the form of hardwired peripherals to support a specific standard (e.g. Brecis's MSP5000 has two on-chip 10/100 Ethernet MACs) or as programmable peripherals that can support multiple protocols (e.g. the Freescale DCP C-5 has two parallel Serial Data Processors that can be micro-coded to support a number of different layer-2 interfaces, like ATM, Ethernet or SONET). This may be viewed as a special case of parallel processing where the data I/O task is handled by the peripherals.

Classification of Peripherals.   We classify interfaces based on usage scenarios and group the interfaces deployed in ASIPs into six classes:

- *Network interfaces* connect to an external network based on fibers, wires or wireless.

- *Memory interfaces* and memory controllers access external memories (e.g. SRAM, SDRAM or compact flash).

- *System extension interfaces* such as external buses or co-processor interfaces extend a system locally (e.g. PCI).

- *Test and debug interfaces* are specialized for test and debug purposes (e.g. JTAG). General purpose interfaces may be used as well, but are accounted in a separate class.

- *Common standard interfaces* are integrated in virtually every SoC system and are system independent. They also serve multiple purposes. So far, we have identified general purpose I/Os and UARTs as class members.

- *Specialized auxiliary interfaces* do not fit in any other class (e.g. synchronous serial interfaces).

1.2   Programmable Hardware Based ASIPs

Programmable hardware based ASIPs tend to be more applicable to data-flow style computation with minimal control flow requirements. Any significant control flow tends to be better supported by ISA-based ASIPs. Elements of this class of ASIPs can be distinguished from each other in terms of the granularity of hardware programmability and time of programming.

1.2.1   Granularity of Programmability.   There are two main points along this axis – fine grain programmability and coarse grain programmability. The building blocks in fine grain programmability are of the order of logic gates – this corresponds to conventional FPGAs. While in some cases there is support for special functions such as fast carry, these fabrics are still intended for general purpose use and it would be hard to classify them as being application specific. Application specific support is provided in coarse grained fabrics where the building blocks are function units that can be customized for the application domain. Likewise, specialized memories (register files as well as larger memory banks) can be customized to provide support for efficient data flow during computation. Examples of this are the Pleiades [238], MorphICs, Quicksilver, and Chameleon platforms for digital signal processing.

1.2.2   Time of Programming.   The two main points along this axis are: Program once before deployment (static reconfiguration) and program possibly multiple times during execution (dynamic reconfiguration). Initial dynamically reconfigurable fabrics required a large number of cycles for the reconfiguration process (of the order of thousands of cycles). Thus, the reconfiguration had to be initiated well in advance. Recent dynamically reconfigurable architectures such as the Chameleon processor provide for single cycle switching between contexts. This is accomplished by storing the different reconfiguration contexts in a distributed memory at each programmable point rather than storing the contexts in a separate unified and remote memory.

These dynamically reconfigurable processors with single cycle context switches can be viewed as instances of VLIW processors with each context being a long (and distributed) instruction. Unlike a VLIW instruction, however, a context is typically executed for a very large number of cycles before being switched and replaced by another context. Nonetheless, we can see a blurring of the line between these dynamically reconfigurable and VLIW ISA processors. This has even prompted a study of the efficiency of exploiting operation level parallelism among these two styles of processors [102]. Going forward, one of the challenges is to view the entire spectrum of ASIPs – from those based on programmable hardware to those based on ISAs in a single unified framework.

2.　A Closer Look into Sub-Spaces

In the preceding section, we have introduced the design axes relevant for a system-level design space of a programmable platform. However, each of these axes may represent an even more complex subspace. In this section, we have a closer look into two exemplary subspaces – the memory and the peripheral subspaces. We do this exercise in order to show the importance of a disciplined approach to inclusively identifying and defining the design space for a following design space exploration. Otherwise, due to the sheer size of the design space, a designer is urged to fall back on previous design experience. This can lead to vastly sub-optimal designs. The importance of design axes is application dependent. Design axes are not necessarily orthogonal to each other. It is unlikely that previous application-specific design decisions are good choices for different application requirements.

The following two subspaces are good examples for design spaces where a comprehensive analysis helps to understand the importance of design axes. Numerous designs already exist that can be used to identify significant design axes. Alternatively, if a space is unknown, techniques from design space pruning, such as sensitivity analysis (see Chapter 5), can be applied to recognize the axes for further investigation.

2.1　Memory Design Space

A description of the design space usually includes several designs that support the required functionality. However, the space should be limited to valid designs that lead to good results under the concerned metric. For the memory design space, the required functionality is the support (i.e. acceleration) of memory communication patterns needed by the application. Examples of these patterns include FIFO/streaming, LIFO, random access and accesses with a range of spatial and temporal locality properties. The common metrics of concern for memory architectures are speed, power, area or a combination of these depending on the application domain. We determine the design space from a customization point of view. We assume that the memory accesses of an application are often dominated by a few patterns. A large improvement can be obtained by providing specialized hardware to optimize these patterns. A similar approach is used in [90]. This idea is comparable to the use of specialized function units for the computation part of the system.

In the following sections we first show the dominant memory access patterns that exist in many applications. Then we illustrate how to select the design space to customize for these patterns. Lastly, we take a look at the cache design space to give a feeling of the size and complexity of the overall design space.

2.1.1   Dominant Memory Access Patterns.   Through some examples, we illustrate how dominant memory access patterns can be accelerated with the addition of specialized memory units. In [90], Grun et al. examine uniprocessor architectures with SPEC95 benchmarks and found that the compress application has 40% of its memory accesses generated by only 19 instruction lines. With the provision of three special memory units, a 39% improvement in memory performance is achieved. For the vocoder SPEC95 application, stream-like access patterns are dominant. A 13% improvement in memory performance is reported with the addition of three special memory units.

We did an initial study on memory references of multiprocessor architectures performing the SPLASH II benchmark. SPLASH II is a shared memory benchmark [247]. In our study, the processors are arranged in an 8x8 array and communicate through a directory based shared memory. A modified version of the Augmint [172] simulator is used. It instruments a compiled parallel program at the assembly level to catch all shared memory access events. During simulation, on every shared memory access, the simulator context swaps to a special event handler. This design allows a simple connection to any user specified back-end for handling shared memory events. In our COMA-flat memory simulator, the network delay in the 2D unidirectional toroidal mesh is calculated as $delay = overhead + hopdelay \cdot hops$. The attraction memory in our COMA for each processor is set to be infinite. The directory uses the simple MSI shared memory protocol. A read miss involves a request to the directory, a request from the directory to a valid cache and then a cache-to-cache transfer between the caches. A regular write-miss from invalid involves a request to the directory, a request from the directory to a valid cache and then a cache-to-cache transfer. The invalidation requests to other shared caches occur simultaneously with the request to the valid cache and the cache-to-cache transfer. An invalidate-write-miss from shared memory involves a request to the directory and invalidate-from-directory to all other shared caches. The directory itself is fully parallelized on each cache line. For each cache line, the directory queues all requests and completes each request in order.

Figure 3.1 shows a summary of the memory access profiles. The horizontal axis identifies the number of code lines of the application sorted by their impact on memory accesses. The vertical axis is the corresponding normalized cumulative memory delay. We used zero delay for the ideal case as comparison, and a 50 cycle overhead and ten cycle delay per hop for the profiled case. The memory delay is thus measured relative to a zero-delay memory.

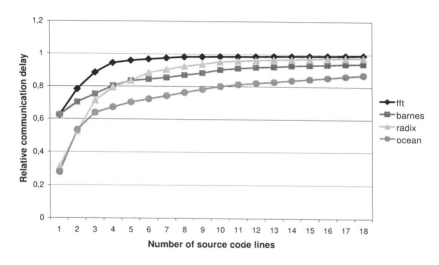

Figure 3.1. Normalized cumulative memory delay vs. number of source Lines for SPLASH II.

Overall, we recognized results that are similar to the studies on uniprocessor systems. The figure shows for the *ocean*, *radix*, *barnes*, and *fft* applications that over 60% of the memory delay is generated from only three lines of source code. Further simulations showed that for *fft*, a 90% improvement in memory performance can be achieved with the addition of two hardware supported shared memory protocols. For *ocean*, a 50% improvement in memory performance can be achieved with the addition of two hardware supported shared memory protocols. Examples of these added protocols are non-binding read and broadcast.

One further example of dominant memory access patterns are streaming applications. Scientific programs and multimedia applications often do not show any temporal locality, but they instead have predictable and regular access patterns. Memory controllers can be designed to exploit this by providing programmable address generators, stream prefetching, and corresponding compiler support [40, 159].

2.1.2   Memory Design Space Selection.   To leverage the potentially large improvement from the hardware speedup of dominant memory access patterns, we adopt the following approach in choosing the design space.

-   The designer classifies the memory access patterns of an application into several compatible groups (e.g. by profiling the application). This identifies the dominant memory access patterns.

-   The designer selects the types of memory hardware units that efficiently support the identified memory access pattern groups (e.g. from a design library).

-   The designer indicates a selection of mapping strategies (i.e. which memory hardware unit can support which memory access pattern group).

-   The mapping strategies combined with the memory hardware choices form the overall memory design space that is chosen by the designer.

To identify and classify the access patterns into compatible groups, the designer can use profiling information and compiler analysis. Access patterns that show similar behavior and can efficiently be supported by a single memory hardware unit are classified into the same group. Dissimilar access patterns that are rarely used can also be grouped together since they are less critical to the overall performance.

By selecting the types of memory hardware units, the designer has to choose those that can efficiently support the memory access patterns groups in the application. The specification can be done by first selecting the general types of memory units such as cache and queue. Component-specific configuration of the unit is then performed. For example, the cache can include the range of associativity, block size, and so on.

Each pattern group must be mapped onto a memory unit. Multiple groups can be mapped onto a single memory unit. Possible mapping strategies can range from all groups mapped onto the same memory unit to each group having its own memory unit. Designers can limit this by choosing an interesting subset of the overall possible mapping strategies. It should be noted that apart from distinguishing memory unit types, we can also have several units of the same type. For example, if there are two memory units used in the mapping strategy and the memory hardware unit types are cache and queue, then there are the following choices. Both units are caches, both units are queues, or one is a cache and the other one is a queue. The complete design space is therefore a combination of the memory hardware types selected and the mapping strategies.

2.1.3   Memory Design Space Complexity.   Although the previous section suggests a straightforward memory design space, the memory design space is actually quite complex. In order to give a better sense of the complexity of the space, we will take a closer look at the cache design subspace. Caching is one of the most commonly used techniques for memory performance improvement. It takes advantage of the fact that many memory access patterns display spatial and temporal locality. It improves average access time by storing data that is more likely to be accessed closer to the computation and in faster memory. Caches can be organized in multiple levels with the base level being the storage memory. The faster and more expensive memory is placed higher in the hierarchy and closer to the processor. Each level of cache can be configured along the axes of set size, block size, associativity, replacement policy, write policy, and so on. Between each level, the connectivity can be configured to a different interconnect topology that can operate based on different protocols.

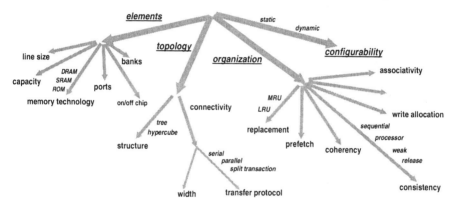

Figure 3.2.   The cache design space.

Figure 3.2 gives a more comprehensive overview of the cache design space. Each arrow represents a design axis. Some major axes are refined into more detailed axes. The major labels name the axes. The smaller labels show a few instances on the axes. In the figure, we can distinguish four main axes. The branch on the left describes the physical characteristics of one cache element (e.g. the memory technology used and the number of implemented read and write ports). The next branch considers the connections between several memory elements including the required communication protocols. The third branch looks at the organization of each cache element and enumerates common caching mechanisms such as miss replacement algorithms. The branch on the right classifies the configuration of a cache by static or dynamic setup. There are more than a dozen design axes for each level of the memory. Many of these axes have wide ranges. For instance, each level's capacity could vary from some kilobytes to several megabytes. Lastly, most of these axes are orthogonal and thus define a large product space.

As an example, we will have a look at the design space definition of a three-level memory hierarchy with the following axes that result in potentially millions of feasible designs.

Level	Parameter	Range	Choices
L1	line size	2, 4, 8	3
	set size/ associativity	1, 2, 3, 4	4
	number of sets	1k - 32k, at 1/2k interval	64
	replacement policy	random, mru	2
L2	line size	8, 16, 32, 64	4
	set size/ associativity	1, 2, 3, 4	4
	number of sets	16k - 256k, at 2k interval	121
Memory	size	128M - 512M, at 64M interval	7

This straightforward example shows us that an exhaustive search of this design space would already require more than twenty million evaluations of different designs. From the definition of the design space, we cannot tell whether some axes have more impact than others. Many axes interact with each other in a nontrivial manner. For example, the effects on conflict misses are related to associativity and line size. Small changes in certain axes can often affect the final design's performance drastically, resulting in a jagged metric curve over the design space. For example, changing the replacement policy from random to MRU can vastly improve the performance of applications that are dominated by streaming data patterns. Thus, we can see that the memory design space can easily become very complex. A systematic approach to exploring the memory design space is needed in order to find an optimal design in a feasible amount of time.

2.2   Peripheral Design Space

Peripherals together with processors and ASIPs, memories, and on-chip interconnection networks are the building blocks for programmable platforms. They can span a large design space as this section will show. Contrary to the memory design subspace, designers are often not aware of design alternatives for peripherals. As a result, peripherals are generally integrated in an ad hoc manner relatively late in the design flow. The goal of this section is to reveal the complexity of the peripheral design space to motivate a disciplined exploration of this space.

To us, peripherals are the communication interfaces that are connected to pins of the chip. Peripherals also include the auxiliary modules that support the interaction between peripherals and processor core(s) (e.g. DMA or interrupt controller). Although the definition may at first seem straightforward, a strict differentiation between platform building block types and peripherals is difficult and not necessarily orthogonal in all cases.

- *Peripherals and on-chip interconnection networks* – The on-chip inter-connection includes all wires and buses that connect modules. It also includes supporting functions, such as bus arbiters and bridges between different on-chip buses. These functions are part of the on-chip inter-connection even though they may implement DMA functions. On the other hand, we consider external bus interfaces to be peripherals because they are connected to pins. They are, however, bridges between on- and off-chip buses and are therefore part of the interconnection network.

- *Peripherals and processors* – Often, processors are combined with hard-ware accelerators that offload the core from computation intensive tasks. If they are just computation-intensive or closely coupled to the process-ing core (aka co-processors), we do not recognize them as a peripheral. However, if an accelerator is associated with a particular communication interface to perform communication related tasks, such as an LCD or a media access controller, we consider them to be part of the I/O subsystem. Even an accelerator that is implemented using a programmable core can be seen as a peripheral.

- *Peripherals and memories* – Peripherals contain local memory. There-fore, they are part of the memory hierarchy in the same way as processors and their register files are. Also, DMA controllers that implement a strat-egy to exchange data between memories are part of the memory hierarchy. Interfaces to off-chip memories and their corresponding controllers are peripherals by our definition, but not internal memories.

As a consequence we recognize that peripherals are an integral part of the platform architecture and should not be handled separately as done in current design flows. Also, their overall impact on the platform's costs suggests a systematic treatment in the design flow.

2.2.1   Impact of Peripherals on Programmable Platforms.   We find peripherals to be a significant share of the die area due to their number and complexity. For instance, with network and media processors, we observe that between 30% and 75% of the area is spent on peripherals.

- IXP1200 – Intel's network processor spends roughly 30% of the die area for only four interfaces.

- Broadcom's BCM4710 – The residential gateway system requires 75% of its die for seven interfaces; only 25% is used for the MIPS core and interconnect.

- Philips Viper/Nexperia™ – The system uses 66% of the die for peripherals and accelerators.

NPUs implement up to 30 different peripherals (ten on average) on a single die that may be instantiated multiple times. In addition, the comparison of areas for embedded processor cores and individual peripherals indicate that particular peripherals can be more complex than a processor core. The complexity greatly depends on the amount of local memory. Cores usually include more local memory than peripherals.

- PCI – The PCI interface in Intel's network processor is as large as four of their RISC-based microengines including instruction and data memory.

- PMM – IBM's physically multiplexed macros are the same size as the complete embedded PowerPC®.

- DSL interface – Broadcom's residential gateway implements a DSL interface (link and physical layer) that is larger than their MIPS CPU including memories.

- 10/100 MACs – Two of the Ethernet interfaces in Broadcom's residential gateway are roughly the same size as the MIPS CPU.

- IEEE 1394 – The IEEE 1394 bus interface in case of Philips' Viper is as large as the integrated MIPS core.

The heterogeneity of peripherals with varying performance requirements has led to several different implementation techniques of peripherals.

2.2.2   Implementation Techniques for Peripherals.   The examination of peripherals in the preceding subsections revealed a tendency towards flexibility in peripherals. This is especially true for the network interfaces of network processors; they are configurable for multiple standards. Other peripherals, such as serial interfaces or memory controllers, can be configured and parameterized as well. We observe three approaches towards flexibility that differ in their degree of configurability apart from hardwired solutions mentioned first.

- *Individual IP* – Every interface is implemented individually and connected to dedicated pins. Although the common case, this leads to a larger number of heterogeneous blocks in the system and to high area consumption. Memory interfaces, for instance, are usually customized for one particular RAM technology.

- *Multiple Macros/pin sharing* – The simplest technique to provide flexibility is the implementation of multiple macros that can be used alternatively in a multiplexed fashion. In this way, pins can be shared among these macros. The obvious drawback of such a solution, however, is the area consumption.

  - The physically multiplexed modules of IBM's PowerNP™ integrate four different network interfaces.

  - Concurrent implementations of USB host and function macros can be found (e.g. in Samsung's residential gateway S3C2510 system). They do, however, not necessarily share pins.

  - Freescale's PowerQUICC™ contains a memory controller with three different modules: A high performance SDRAM interface module, a more power efficient lower performance module (e.g. without burst mode) and three parameterizable modules.

- *Parameterizable Peripherals* – Virtually all peripherals can be adjusted to some extent by specifying a well-defined set of parameters. The degree to which a peripheral is parameterizable (the set and range of parameters) varies with the interface. Here are a few examples:

  - UARTs – Serial interfaces can be configured, for instance, for different transmission speeds and the use of hardware support for protocols (e.g. Xon/Xoff).

  - Memory Controller – Timing parameters and organization of the external memory modules can be set up at boot time.

  - PCI Interface – Some PCI interfaces support multiple versions that differ in bus frequency and width.

  - IBM's Physical MAC Multiplexer (PMM) – The selection signal for multiple (fixed) macros is also a parameter.

  Since such parameters are configured (i.e. programmed) by the processor core and the number of parameters may be substantial, these interfaces are sometimes falsely considered to be programmable.

- *Programmable Interfaces* – In some peripherals, parts of the functions are specified in software. Such programs are executed by dedicated processors [207]. The capabilities of these processors as well as their association to particular interfaces vary:

  - Freescale's Serial Data Processors (SDPs) – These micro-programmable engines handle physical layer tasks of network interfaces such as data parsing, extraction, insertion, deletion, validation and coding. Each network interface uses two SDPs and supports multiple communication protocols.

- Intel's Network Processing Engines (NPE) – The IXP 4xx systems implement these programmable accelerators to support particular network interfaces. These engines are specialized to the needs of an interface, but are based on the same core architecture. This helps to offload computation intensive tasks from the CPU.

- Freescale's Communication Processor Module (CPM) – The CPM contains a processor core that is shared among multiple interfaces and only executes communication protocol layer tasks.

In addition to these approaches, we also find orthogonal reuse techniques that can be combined with any of the approaches above but actually do not require any configurability.

- *Multiple Channels* – Multi-channel peripherals handle multiple (communication) contexts simultaneously. They exploit the fact that some communications (e.g. time slots in a time division multiplexed link) do not require the full performance of a peripheral. Some of our examples also support multiple protocols:

  - TI's multi-channel serial interfaces (MCSIs) – In TI's OMAP^{TM} processors, MCSIs are used that are configurable (e.g. in clock frequency, master/slave mode, frame structures or word lengths). Their buffered version (McBSP) allow continuous data streams for a number of different communication protocols (T1/E1, AC97, I2S, SPI).

  - Freescale's multi-channel controllers (MCCs) – Each MCC supports up to 128 independent TDM channels (HDLC, transparent or SS7).

- *Use of generic interfaces* – An additional approach to increase reuse and flexibility is the use of generic interfaces instead of specialized ones. We observe three different applications:

  - Wireless application processors use interfaces, such as UARTs, rather than proprietary solutions (e.g. to connect to companion chips (e.g. Bluetooth^{TM} PHY)).

  - Almost every system provides a number of general purpose I/Os that can be used under software control for any I/O function at the expense of CPU performance.

– Some systems implement generic interfaces, such as the serial management interface in IBM's network processor, that may require additional external logic to support less common usage scenarios.

Programmable platforms contain processor cores and ASIPs, memory hierarchies, interconnection networks, and peripherals as main building blocks. During the development of a platform, a lot of emphasis is put on the customized design of the first three aspects. Peripherals are often neglected and their complexity is underestimated. This is mainly due to the general perception that such building blocks constitute standard intellectual property that can be acquired as needed. These blocks are therefore integrated separately and relatively late during the design process. This section has shown that the peripheral design space is complex. The costs for peripherals can be a significant part of the overall cost of the platform. We therefore believe that the peripheral design space is a good example that underpins the necessity of a disciplined approach to design space exploration of programmable platforms.

3. Network Processing Design Space

The domain of network processing has been a particularly active area of research for the last couple of years. The architectural space of network processors serves as an example in this section to reveal the complexity of application-specific programmable platforms. In the preceding sections, we have introduced design axes for programmable platforms in general. In this section, we reveal by a comprehensive overview of the design axes for network processors (NPUs) how many design axes are actually significant in real word designs. This procedure is required to gain a thorough understanding of the design space to guide the development of new designs in a disciplined way. The NPU design space is quite diverse and uncovers only a few common properties among designs. This diversity is rather based on prior design experience in the respective companies than on a systematic characterization of the design space. We believe that a comprehensive review of the design space should be a major design step before any particular design point is implemented in order to allow a systematic exploration of the design space. The ASIP design space is complex and poorly suited to exploration through human intuition alone.

The NPU space example given in this section is based on a comprehensive review of existing designs to determine the sensitivity of the design space to particular design axes. If the design space is rather unknown and only a few designs exist, methods from design space pruning and high-level evaluation can be reused to determine interesting regions of the design space. These hierarchical and early design space exploration techniques have their own place in this book and are further described in Chapter 5.

3.1   Parallel Processing

On the coarse grained processing element level, we find different approaches to deal with and exploit concurrency. First, a field of homogeneous processors is often used to perform the main data plane processing functions. Figure 3.3 compares network processors according to their throughput class and number of cores used. The throughput values are best case scenarios advertised by the manufactures.

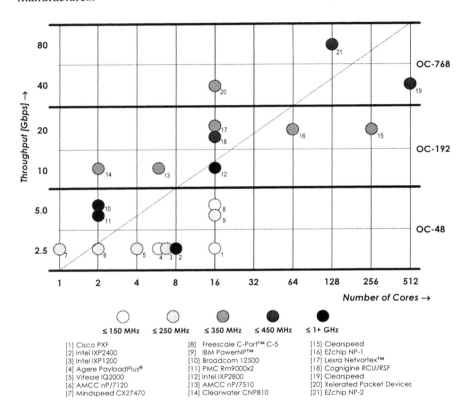

Figure 3.3.   Number of cores and throughput class for different NPUs.

We can see that developers have used quite a different number of processing cores within the same performance class, and the same number of cores can span a wide range of achievable throughput.

Apart from using cores concurrently, several threads of execution can be mapped onto the same core. Cores in NPUs often offer hardware support for this mode of operation (e.g. fast context switches by using separate register banks in order to virtually hide access latencies to shared and/or slow resources such as memories). From an application point of view, these threads virtually execute concurrently. Figure 3.4 compares NPUs in terms of number of threads supported per processing core.

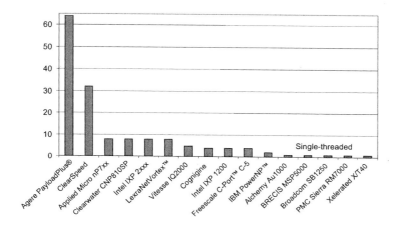

Figure 3.4. Hardware thread support among NPUs.

The single threaded variants often target the low end access network segment or control plane processing. Looking at NPUs targeted at edge and core networks, we recognize that the number of threads varies by more than one order of magnitude. It is not clear from Figure 3.3 whether more processing cores or more threads per core make sense to exploit concurrency.

Finally, when we look at the interconnect topology of network processors, we see that some manufactures prefer a pipelined organization of the processing cores, whereas others favor a pool of processors. There are also intermediate solutions such as a pool of pipelines. NPU processing core topologies are compared in Figure 3.5.

Instruction level parallelism is exploited by some of the NPUs by supporting functional units on a single processing core with multiple issue slots. Most of the NPUs only support an issue width of at most two. Figure 3.6 gives an overview of the issue width per processing core versus the number of processing cores.

Bit level parallelism is exploited by NPUs using dedicated function building blocks and instructions. Examples are the concurrent extraction of header fields and the independent processing of these fields. Arithmetic instructions in NPUs can use bit masks for this purpose, whereas traditional general purpose instruction sets would require several shift and byte/word mask operations to do the same task.

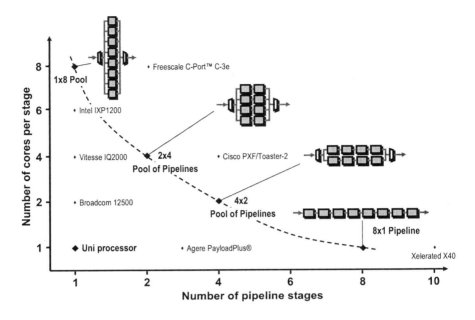

Figure 3.5.   Processing core topologies.

3.2   Special Purpose Hardware

The computational kernels are prime candidates for hardware acceleration on an NPU. The key parameters in designing special purpose hardware blocks are functionality, interface, and size. That is, whether the hardware implementation fits within the pipeline. The two most common approaches to providing specialized hardware are:

- Co-processors – may store state and have direct access to memories and buses.

- Special functional units – usually stateless and within a pipeline and only write back to registers.

Most network processors employ both of these techniques extensively for accelerating lookup, queue management, pattern matching, checksum/CRC computation, and encryption/authentication. Figure 3.7 gives an overview of how specialized functionality is implemented in different NPUs.

This comparison shows the diversity among different network processors. For example, IBM and Freescale have co-processors for most or all packet processing kernels, while Cognigine relies solely on their reconfigurable functional units. EZchip has entire processors devoted to pattern matching, lookup,

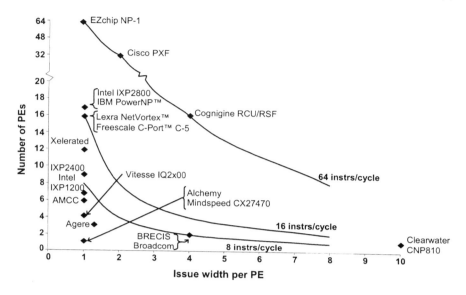

Figure 3.6. Issue Width per processing element versus number of processing elements.

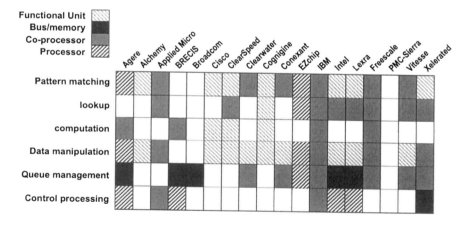

Figure 3.7. Specialized hardware employed by NPUs.

data manipulation, and queue management. On the other hand, PMC-Sierra's network processor has no specialized hardware. Broadcom and Alchemy have very little. A number of processors have interesting mixes of functional units, memory features, co-processors, and processors for various tasks. Agere's PayloadPlus® system uses a special processor for pattern matching and data manipulation, a co-processor for checksum/CRC computation and has memory features for queue management. Vitesse and Xelerated Packet Devices shy away from special bus or memory features and simply use a mix of coprocessors and function units. Intel and Lexra also include special memory and bus features and have a dedicated processor for the control plane.

This diversity of designs can be explained by the affinity of manufactures to reuse existing building blocks as often as possible rather than evaluating previous design decisions under the new constraints of the application domain.

3.3   Memory Architectures

Memory architectures used by NPUs are typically heterogeneous. They vary from slow and large external DRAM to fast on-chip SRAM-based scratchpad. Technology-wise, we can thus find all flavors of registers, SRAMs, DRAMs, off- and on-chip memory variants in a single NPU. In order to increase throughput, several memory buses might even be used concurrently. The different memory technologies are more thoroughly discussed in the following peripherals subsection since they require dedicated interfaces.

Memory organization is usually flat (i.e. there are no caches), but the programmer can explicitly choose fast or slow memory. A natural split of the packet content therefore often favors storing packet headers in the fast memory portion and the payload in the slow memory. In this way, it is assumed that most of the processing will appear on the header fields and not on the packet content. As shown in Figure 3.7, some NPUs offer accelerators to offload certain memory management functions, such as keeping track of FIFO organized queues for packet payload.

Apart from the main fast and slow memory areas that are shared among all processing cores, we also find buffer memory distributed over the NPU microarchitecture. These buffers are required to couple different clock domains (e.g. a physical link from a processing core). In the Intel IXP, for instance, examples are the transmit and receive FIFOs and the FIFO buffers on the MAC layer at the IX bus. Other specialized memories include content addressable memories (CAMs) that can be used for implementing address lookups and classification.

3.4   On-chip Communication Mechanisms

In general, on-chip communication mechanisms are tightly related to the configuration of the processing cores. For NPUs with pipelined processing elements, most communication architectures are point-to-point between processing elements, co-processors, memory or peripherals. NPUs with symmetric processing core configurations often have full connectivity with multiple buses. For example, Freescale's C-Port^{TM} C-5 DCP has three buses with a peak bandwidth of 60Gbps that connects 16 channel processors and five co-processors. However, this solution has limited scalability. The use of on-chip packet switching networks is emerging as a scalable alternative to buses.

3.5 Peripherals

In this subsection, we examine a representative set of network processor architectures from Intel, Freescale, Broadcom, and IBM and compare them with respect to their interfaces. We first classify NPU interfaces according to their purpose.

3.5.1 Interfaces Deployed with NPUs. Network processors are often deployed on line cards in network core and access equipment (see Section 3). Line cards manage the connection of the equipment to the physical network. They are connected to a backplane that allows packets to be distributed to other line cards or to control processors. Due to this usage, NPUs require a number of dedicated interfaces: *Network interfaces* connect the processor to the physical network, a *switch fabric interface* accesses the backplane, a *control plane interface* connects to an external processor that handles the control plane and also maintains the NPU and its peers on the card, *memory interfaces* handle packets and table lookup data and *co-processor interfaces* are used for classification, quality-of-service and cryptography accelerators.

While some of these interfaces (e.g. memory and network interfaces) are well covered and documented by mature I/O standards, others, especially the switch fabric interface, still lack sufficient standardization [63].

The interfaces deployed by the examined network processors are listed in Table 3.1. They are grouped according to their classes. The table confirms the fact that network processors deploy a large number of quite diverse interfaces. The table also reveals the diversity in memory and network interfaces among the processors. In the following, we examine these interfaces more closely.

3.5.2 Network Interfaces. Network processors implement two sets of peripherals to connect to the physical network: Ethernet and ATM. These interfaces usually perform OSI Layer 2 functions – MAC in case of Ethernet and SAR for ATM. The physical layer (layer one, PHY layer), which is normally provided externally, is connected via one of the following three interfaces:

- The *Media Independent Interface* (MII) or one of its variants (RMII, SMII) is commonly used for Ethernet and virtually supported by every NPU. These interfaces come in three performance classes: 10/100 Mb/s (MII), 1 Gb/s (GMII) and 10 Gb/s (XGMII).

- The *Utopia interface* is used for ATM environments to connect external PHY devices. Utopia has three throughput classes: 155 Mb/s (L1), 622

Table 3.1.   Network processors and their I/O interfaces.

SoC Interface	Intel IXP 12/24/2800	Freescale C-Port™ C-5	IBM PowerNP™	Broadcom BCM 1250
network interfaces				
number of interfaces	1	16x CP[a] + 1 SW[b]	4x PMM + 2 SW[b]	3
IX-Bus	(yes)	–		–
Utopia	yes	yes	(yes)	–
CSIX	yes	yes	yes	–
SPI/POS-PHY	yes	unknown	yes	–
MII/GMII/TBI	–	yes	yes	yes
Flexbus	–	–	yes	–
memory interfaces				
SRAM	1x /2xQDR™	2xZBT®	2x QDR™/LA-1/ZBT®	
SDRAM	1x /DDR/3xRDRAM®	SDRAM (PC100)	8x(DDR)	DDR
extension interfaces				
PCI	yes	yes	yes	yes
(NPLA-1)	–	–	(yes)	–
Hypertransport	–	–	–	yes
test and debug				
JTAG	1x	1x	1x	1x
common				
Core/Type	yes/ARM	no	yes/PowerPC	2x/MIPS
GPIO	4-8x		3x	yes
UART	1x	(MDIO)	no	2x
auxiliary				
MDIO	0/(yes)	yes	(SPM)	–
Serial boot ROM	1x	1x	(SPM)	2x(SMB)

[a] one per channel processor.
[b] switch interface.

Mb/s (L2) and 3.2 Gb/s (L3). For higher throughput the SPI interfaces are used.

■ *System-packet interfaces* (SPI), which are similar to Utopia, are used for ATM and packet-over-SONET (POS) environments to interface to external framers. SPI/POS-PHY classes range from 622 Mb/s to 40 Gb/s (SPI-5).

In order to manage the PHY modules connected to the networking interfaces, an auxiliary PHY management interface (MDIO) is provided by all NPUs. IBM's SPM module, for instance, requires an external FPGA to do so. In this way, the SPM module can support different interface types.

In addition to Utopia and SPI, network processors that dedicate peripherals for interfacing to an external switch fabric use one of the following options:

- The *common switch interface* (CSIX-L1) is a relatively wide interface that needs 64 data pins at 250 MHz to support 10Gb/s and in-band flow control.

- The *network processor streaming interface* (NPSI) is based on 16 LVDS data pins at 311-650 MHz and two to four additional flow control bits.

Proprietary interfaces become more and more obsolete and are successively replaced by standardized interfaces. Intel, for instance, decided to replace the IX Bus in its second generation NPUs by SPI/Utopia interfaces.

3.5.3   System Extension Interfaces.   The system extension interfaces of network processors fall into two categories: control plane and co-processor interfaces. In theory, both types can be used for the same functionality. The control plane interface, however, is more likely to be generic and well established. The co-processor interface, on the other hand, might be optimized for low latency interactions. In practice, we find only one standard control plane interface commonly used:

- The *peripheral component interface* (PCI) provides a local bus interface with a peak bandwidth ranging from 1 Gb/s to 4.2 Gb/s depending on its version. The newer PCI-X® 1.0 and PCI-X® 2.0 specifications provide point-to-point peak bandwidth between 8.5 and 34 Gb/s at {133, 266, 533} MHz x 64 bit, but are not (yet) integrated into products.

Some network processors, such as AMCC's, integrate proprietary interfaces to connect to peer and control processors. These interfaces may be replaced in the future as more standardized alternatives to PCI (e.g. RapidIO and Hypertransport™) become available.

Besides proprietary or network interfaces (SPI), co-processors are often memory mapped and connected via modified memory interfaces. For this reason, the Network Processor Forum specified the NPLA-1 interface:

- The *look aside interface* is a model based on a synchronous DDR SRAM interface and provides a bandwidth of 6.4 Gbit/s per direction at 2x 18 bit x 200 Mhz.

3.5.4   Memory Interfaces.   Although NPUs often integrate considerable on-chip memories, they additionally need at least two different kinds of external memories: A large packet memory and fast memory for auxiliary data. Our NPU examples deploy the most recent memory interfaces, often in multiple instances, to meet their memory bandwidth requirements:

- *Double data rate* (DDR) SDRAM supports data transfers on both edges of each clock cycle, effectively doubling the data throughput of the memory device. The ones used in our examples provide between 17 and 19.2 Gb/s at 64 bit x {133, 150} MHz.

- *Rambus' pipelined* DRAM (RDRAM®) uses a 16 bit-wide memory bus. Command and data are transferred in multiple cycles across the bus (packet-based protocol). Components are available ranging from 800MHz to 1.2GHz and provide 12.8 Gb/s - 19.2 Gb/s peak memory bandwidth.

- *Zero Bus Turnaround* (ZBT®) SRAM do not need turnaround bus cycles when switching between read and write cycles. This is beneficial in applications with many random, interleaved read and write operations, such as accessing routing and lookup tables (e.g. Freescale's NPU deploys a 8.5 Gb/s at 64 bit x 133MHz interface).

- The *Quad Datarate* (QDR™) SRAM uses two separate memory ports for read and write that operate independently with DDR technology; thus effectively doubling the memory bandwidth compared to DDR. Since both ports are separated, bus turnaround is avoided. The IXP 2800 implements a total of 4x 12.8 Gb/s at 32 bit x 200 MHz on memory based on QDR™.

These memory interfaces are standard interfaces. Memory controllers embedded in NPUs are however more specialized compared with their general-purpose counterparts. They at least need to support multiple processing elements, which may access a memory concurrently (e.g. by providing individual command queues). In addition, application-specific knowledge is exploited to tailor controllers to the specific usage. In Freescale's example, the queue management, buffer management, and table lookup units are specialized to their specific data structures and access functions.

3.5.5   Other Interfaces.   NPUs that implement a standard control processor also implement the common GPIO and UART interfaces.

This survey has demonstrated that complex interfaces are deployed at all key positions of NPU micro-architectures. We therefore believe that their impact on the system's cost and reusability has spuriously been disregarded in current design frameworks. Currently evolving standards for high speed interconnects, such as PCI-Express™, RapidIO and Hypertransport™, will make matters worse [207].

4.   Conclusion

In this chapter, we have shown that the design space for programmable platforms is complex and composed of even more complex subspaces. We believe that a comprehensive review of the design space should be a major design methodology element before any particular design point is implemented. The platform design space is complex and poorly suited to exploration through human intuition alone. We have inclusively identified design axes that play a significant role in ASIP and programmable platform design. Depending on the chosen application domain, the set of design axes for further design space exploration can be identified by applying:

- A comprehensive analysis of existing designs.
- Design space pruning techniques.

For the former case, we have shown, using the example of network processors, how a comprehensive analysis of existing designs helped us to find the most prominent design choices in this application domain.

In the latter case, the first element of our methodology is very helpful. Since benchmarks have been selected based on a careful analysis of the application domain, they can be used to identify corner cases of the design space (e.g. by using hypothetical system architectures and analytical models or simulation). Early design space exploration techniques mention here have their own place in this book and are further described in Chapter 5.

Ideally, in order not to be bound by existing design choices that may not cover the whole design space of interest, both principles should be applied to some extent so that the design space is restricted systematically.

Chapter 4

EFFICIENTLY DESCRIBING AND EVALUATING THE ASIPS

Scott Weber[1], Yujia Jin[1], Matthias Gries[1], Christian Sauer[2], Matthew Moskewicz[1]

[1] *University of California at Berkeley*
Electronics Research Laboratory

[2] *Infineon Technologies*
Corporate Research
Munich, Germany

After having identified the design space of an application-specific programmable platform systematically, the question arises how to develop the platform in a disciplined and efficient way. In this chapter, we review how existing approaches deal with the design of application-specific parts and derive principles that allow a more efficient and possibly more natural design flow to correct implementations. Our own design flow named Tipi realizes these principles. The description of Tipi's specifics is a considerable part of this chapter. We illustrate the usefulness of our tool flow with several design examples.

In the following section, we review the current state-of-the-art of designing programmable cores. In Section 2, we characterize mandatory properties for a disciplined development of ASIPs and show how Tipi implements these principles. Section 3 describes in detail what techniques are used in Tipi for realizing a disciplined correct-by-construction design flow. Case studies showing the effectiveness of Tipi's approach to ASIP design are discussed in Section 4. In Sections 5 to 7, we show how the Tipi design flow can be used for design sub-spaces as well; for instance, the memory design space and interconnects. Tipi thus provides all necessary functionality for describing and evaluating programmable platforms efficiently.

1. Best Practice in ASIP Design

Traditionally, the number of programmable architecture design starts has been few. As a result, little commercial tool support exists for this market. But with the increasing replacement of application-specific integrated circuits (ASICs) by the application-specific instruction-set processors (ASIPs), it seems natural that electronic design automation should support such systems. These designs require a suite of development tools including a simulator, assembler, and compiler. To actually realize the hardware design, a netlist in a traditional hardware description language (HDL) must be generated from a description of the architecture.

1.1 Architecture Description Languages (ADLs)

A variety of architectural description languages (ADLs) [195, 233] have been proposed to assist in the design of programmable architectures. A common aspect of existing ADLs is to restrict the scope of architectures supported in order to manage the complexity of the associated tool generation. We feel that these restrictions will force the designers of emerging irregular architectures, such as the Intel IXP1200 [108], Broadcom Calisto™ [174], and Xilinx Virtex II™ Pro [249] to continue the use of traditional design practices that involve the manual construction of simulators, assemblers, compilers, and HDL models.

1.1.1 Behavioral ADLs. So far all ADLs must make trade-offs among ease-of-expression, breadth of architectures supported, and the quality of the generated tools. Behavioral ADLs (nML, ISDL, and CSDL) allow designers to describe the instruction semantics to create a programmer's view of the architecture. In fact, Target Compiler Technologies[1] has commercialized the use of nML in its Chess/Checkers [138] environment. However, the lack of sufficient structural details in behavioral ADLs makes it difficult to extract an efficient micro-architecture.

1.1.2 Mixed ADLs. To attack the limitation of behavioral ADLs, mixed ADLs (FlexWare, Maril, HMDES, TDL, EXPRESSION, and LISA) were developed to enable designers to describe the instruction semantics, the micro-architectural structure, and a mapping between the two. In fact, companies such as AXYS[2] and LISATek™/CoWare[3] are providing high performance cycle-accurate simulators and system integration tools based on custom processors described with LISA [186]. However, these ADLs, as well as the behavioral

---

[1] *http://www.retarget.com*
[2] *http://www.axysdesign.com*
[3] *http://www.coware.com*

ADLs, lack support for designs outside a defined family of architectures, and system integration is performed in an informal manner.

1.1.3   Stylized HDLs.   Stylized HDLs such as the structural ADL MI-MOLA [150] capture the concept of instructions albeit with the requirements that the instructions are single-cycle microcoded operations. The MIMOLA effort demonstrates that it is possible to extract instructions [148], generate a compiler and a simulator, and synthesize a design from a single low-level description. Tools supporting MIMOLA, however, assume certain uses of architecture features, such as a program counter, a register file, instruction memory, and single-cycle operations.

1.2   Hardware Description Languages (HDLs)

One way to model architectures outside the scope of these ADLs is to use an HDL such as Verilog®, VHDL, or SystemC™. These languages support any architecture, but lack the notion of operations. This manual construction is time consuming and error-prone. If you agree that ASIP architectures will be increasingly diverse and irregular, then it is imperative to provide a better methodology to enable their design.

2.   Recipe for Developing ASIPs

In this section we derive a set of design principles that should be followed in order to implement a disciplined and efficient design flow for ASIPs. The goal is to ease the development of micro-architectures and supporting tools efficiently, such as simulators and assemblers, while offering a path to implementations in hardware. Manual intervention by the designer should be reduced to a minimum in order to avoid error-prone adaptations among different aspects of the design. We would like to note that the discussed ADLs already support a subset of these principles and thus simplify ASIP design. However, so far no tool flow has been developed that supports all principles seamlessly, particularly for irregular ASIPs.

2.1   Design Principles

The following rule set defines a framework for efficiently designing and implementing integrated circuits. The Tipi design flow follows these rules and enables the fast development of application-specific programmable parts, particularly defined by irregular micro-architectures.

2.1.1   Strict Models.   The underlying models, which describe the information from design entry, must be strict in the sense that they should only allow valid, e.g. synthesizable, designs. These formally defined models ease verification and correctness of the design.

2.1.2   Separation of Concerns.   In order to discourage manual tuning of a design, different aspects of the design should be handled independently. For instance, it is error-prone to describe the datapath and the corresponding control logic together since data and control flow aspects would be mixed in a single description. When modifying the datapath in this mixed description, the designer has to continuously monitor and adapt the control part as well, which is confusing and may lead to inconsistent designs.

2.1.3   Correct by Construction.   The correct-by-construction principle brings strict models and the separation of concerns together to enable an efficient design flow. It means, for instance, that different aspects of a design are kept consistent in underlying formal models without any interference of the designer. That means, the designer can concentrate on the actual design aspect under development while the internal model is maintained and adapted transparently. Correct-by-construction also implies that certain supplied tools for evaluating and deploying the platform, such as a simulator and an assembler, are generated correctly from formal models rather than manually written by designers. In this way, verifying the behavior of these tools against the actual design can be avoided since the strict models and the generated tools are always and provably consistent.

2.2   Tipi's Implementation of the Design Principles

Existing ADL-based approaches implement some of the design principles well for a limited set of architectures. We would like to emphasize the fact that the Tipi design flow enables a wider set of architectures to be described. In addition, our more natural abstractions from the underlying micro-architecture simplify the overall design process.

2.2.1   Correct by Construction Design.   Tipi uses one single internal representation of the design, from which all supporting tools, such as simulators and assemblers, and HDL descriptions are derived. Tipi shares this technique with existing ADL-based tool flows. Moreover, different design aspects are kept consistent in the model transparently for the designer.

2.2.2   Multi-View Design.   In Tipi, the designer can choose between different views of the design. This means, different concerns/ aspects of the design are represented by different views, e.g. datapath, control logic aspects, and the memory subsystem. These views are automatically extracted from the single internal representation. The internal model is always kept consistent automatically if the designer modifies one aspect of the design.

2.2.3   Operation-Level Design.   Tipi introduces a new abstraction called operation-level design. We believe that a designer of a programmable application-specific part should only be concerned with the design of the datapath of the design. The necessary control logic can be seen as an overhead of the overall design process. Since an application-specific programmable part need not be compatible with preceding designs, the instruction set can be chosen specifically for the targeted application domain. In existing ADL-based tool flows, the designer still has to specify the instruction encoding and possibly the control logic explicitly. This is not necessary in Tipi. From the description of the datapath, all possible operations are systematically extracted that the datapath supports. These operations can be used to define a customized instruction set. At all steps, Tipi guarantees that the underlying micro-architecture implements the instruction set. All further tools, such as simulator and assembler, are extracted accordingly.

The correct-by-construction, multi-view operation-level design approach implemented by Tipi will be described in detail in the following section.

3.   The Tipi Design Flow

Architectural description languages are increasingly being used to define architectures and to generate key elements of their programming environments. A new generation of application-specific instruction processors is emerging that does not fit well in an instruction-set architecture paradigm. In this section, we propose a methodology that supports the design of a wider range of programmable architectures than existing architecture description languages. We achieve this through a flexible framework that is centered on a constraint-based functional hardware description language that abstracts computation using the notion of primitive *operations*. Using multiple *views* acting on a common model, our methodology ensures consistency between architectural design, tool generation, and hardware implementation. We term our methodology *multi-view operation-level design*.

Tipi's overall design flow is displayed in Figure 4.1. The shaded area is the focus of the following section, which explains the design entry process. The central block in this figure also constitutes the central underlying idea of Tipi: To automatically extract operations supported by the datapath to enable correct-by-construction instruction sets and implementations. The remaining steps for generating fast simulators and implementations are described in detail in Section 3.2.

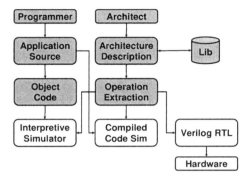

Figure 4.1. Tipi design flow.

3.1   Multi-view Operation-level Design – Supporting the Design of Irregular ASIPs

To support this new generation of ASIP architectures, we propose a component abstraction above the register-transfer level (RTL) abstraction that allows the design of any programmable architecture that can be described as an arbitrary collection of data paths which can be configured. RISC datapaths, systolic arrays, special purpose co-processors, and FPGA look-up tables all fall within this category. From each datapath description, we automatically extract all operations that the datapath can execute. From these extracted operations, we can generate the tools required for the programmable architecture design. These tools include high-speed cycle-accurate simulators and an assembler. It is important to note that we can produce these tools without committing to a particular control strategy (e.g. horizontal microcode, traditional RISC controller, and so on). The design can be simulated with control abstracted as a simple trace of the operations to be executed. Ultimately, a controller that implements the control strategy must be chosen so that a synthesizable RTL hardware model can be produced for the complete design.

At the core of our methodology is a common model with well-defined semantics. The model contains information about the structure of the architecture and the supported operations. Multiple *views* (synonymous with abstractions) are then developed on top of the model to encapsulate and present the information required by each facet of the design process. Practically, a view consists of a set of tools that operate on a model to extract some particular information or perform some transformation. Views have the invariant that they must respect all of the semantics of the model, and never infer anything that is not present in the model. This implies that once a view is (correctly) written, the outputs of that view will be correct by construction with respect to any model. Thus, the multiple views enable designers such as the simulator writer, assembly writer, and compiler writer to work with the design from differing but always consistent perspectives. The key to achieving semantic consistency over such a broad range of architectures in our methodology is an operation-level description that captures both structural and behavioral aspects of the design while orthogonalizing control aspects.

This subsection is organized as follows. In the next subsection, we review prior work on ADLs. In Section 3.1.2, we discuss multi-view design. In Section 3.1.3, we use a small example to illustrate how one designs a micro-architecture in the architecture view. Section 3.1.4 discusses how the operation view is generated. Section 3.1.5 discusses how the high-speed cycle-accurate simulator is generated. Section 3.1.6 discusses the assembly view. Section 3.1.7 discusses the hardware view. A case study is presented in Section 4.

3.1.1   Related Work.   Within the last decade, a variety of ADLs (references to the ADLs named here can be found in the surveys [195, 233]) have been introduced to assist in the development of new architectures.

The existing ADLs have certainly enabled the use of a broader scope of programmable architectures in designs. However, ADL developers are faced with the dilemma that they want to support a wider range of architectures without sacrificing generated tool quality. We address this dilemma by developing an operation-level abstraction capable of modeling any programmable architecture. We then formalize the activity of generating tools and HDL models from this abstraction as a process of restrictions and invariant preserving transformations. The range of architectures that we support is then no longer a function of the ADL but of the breadth of the available views.

3.1.2   Multiple Views.   Traditionally, an HDL is used to describe a micro-architecture. However, HDLs lack semantics to make a formal distinction between the control logic and the datapath. Architects instead make this distinction by specifying the instructions and the corresponding control logic manually. Validating the correctness of these instructions requires extensive simulation and is error prone. To alleviate the pain of verification in our approach, we have developed an operation-level abstraction that separates control from data, thus enabling the exportation of the architecture as a set of operations.

Abstracting an architecture as a collection of operations is a key enabler for our multi-view methodology. The abstraction gives us the model from which the simulator, assembler, and compiler *views* can be developed. Using these *views*, the simulator, assembler, and compiler can then be automatically generated. Furthermore, the underlying model allows us to develop these multiple views without requiring a single cumbersome syntax for all views. We feel that no single syntax is natural for architects, compiler designers, and simulator writers because each is concerned with different facets of the design.

Our methodology dictates that views must work only with semantics explicit in the model. We do not guarantee that a view will succeed in manipulating the model. It is up to the designer of a view to ensure that any transformation that a view applies to the model has the intended effect, and that any outputs of the view represent correct extractions of information in the model. The issue of designing and verifying the views themselves is outside the scope of this section, but there are many alternatives, including standard random simulation as well as formal verification techniques similar to those used to verify that multiple specifications of a single processor conform to some property [37].

To illustrate the benefits and limitations of the multi-view methodology, consider the creation of a compiler view. The intent of a compiler view is to provide a mean to automatically generate a compiler given an architecture. Because our underlying model is quite descriptive, it is possible to design an architecture outside the scope of a compiler view. For example, a compiler view that only understands non-clustered architectures may fail if the architecture has a split register file. To solve this problem, either the architecture view is restricted or the compiler view is enhanced. It is exactly this appropriate defining, partitioning, and restricting of views that is the cornerstone to a successful framework.

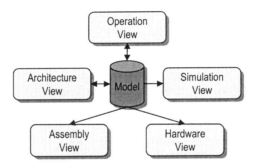

Figure 4.2. Multiple views.

To demonstrate the methodology, we propose an initial framework composed of the five views shown in Figure 4.2. Designers can compose hardware blocks in the architecture view. The computational functions that can be performed by the architecture are extracted and presented by the operation view. The simulation view uses information from the architecture and operation views to generate a high-speed cycle-accurate simulator. In order to program the architecture, a simple traditional assembler is generated in the assembly view. Finally, a synthesizable RTL hardware model is generated in the hardware view, thus enabling a realization of the architecture. These views represent an introduction to a framework which will also include a compiler view.

3.1.3   Architecture View.   The architecture view provides the designer the ability to describe the datapath. In our architecture view, datapaths are described by connecting *actors* with *signals* between *ports*. The term *color* refers to an abstract property used to distinguish signals. Rather than encoding signals as a collection of bits (as in a traditional HDL), we instead abstract the interesting properties of signals by choosing appropriate colors. The coloring of a signal subsumes the traditional notions of a signal of having a "type", a constant value, or a symbolic value. Generally, a signal has a set of possible colors determined by its context. There is a color, "not present", which corresponds to a logical "X" in a three-valued logic. A signal has this color when it contains no information. Almost all signals have the "not present" color in their set of possible colors. Signals such as the control inputs to multiplexers are generally abstracted as enumerations of constants because all possible values of these signals must be distinguishable during analysis.

Colors can be grouped together into sets (hereafter referred to as color-sets). Besides equality, which is defined over all colors, color-sets may have additional axiomatic relations defined. For example, the bit vector color-set has the standard logical and arithmetic functions that one would expect in an HDL. The colors in a color-set not only include constants (expressed as "color-set.color"), but also symbolic variables (expressed as "color-set.varname"). For example, each source in a circuit will introduce a fresh symbolic color. The signal bound to each source will have only this color in its set of possible colors (technically, because sources can be unused, "not present" is also a possible color for source signals). Additional symbolic colors are introduced to represent all possible computation by actors. All such computation by actors occurs within *firing rules* which define the axiomatic relations between inputs and outputs. Firing rules can only be active when certain color constraints on the signals bound to the ports of the actor are met. Note that this is not the same as a simple color-set constraint, because it can also test against specific colors. Additionally, actors specify a validity constraint using first-order logic over the "activeness" of their firing rules. Typically, this constraint specifies that at least one firing rule is active, but more sophisticated constraints are possible and sometimes necessary. In our current system, we have defined the three color-sets as shown in Table 4.1.

Table 4.1.   Color-set definitions.

Color-set	Description
X	Set containing only "not present" color
enum	Set of colors for abstracting enumerations
bit	Set of colors for abstracting bit vectors

Examples in Figure 4.3 illustrate how actors are defined in our language. For each firing rule, the color constraints are in parentheses after the name of the rule, and the axiomatic relations are contained in the curly braces. The first actor is a demux which has two rules for passing the input to the appropriate output port, and one rule that does nothing. The second and third actors represent an incrementer and decrementer, respectively. The final actor is a mux. Note that the color constraints for the mux and demux match against a specific color on the control signal, but only check color-set membership on the other signals. This distinction is typical of the difference between what are traditionally considered control and data signals. When analyzing different configurations of the hardware, control signals tend to have constant values whereas data signals tend to remain symbolic variables. However, the framework makes no formal distinction between control and data signals.

Although not shown in Figure 4.3, two primitive state actors, the *register* and *flip-flop*, are also defined. A *register* holds a value written to it until the value is overwritten. A *flip-flop* holds the value written to it for one cycle. Both of these state actors are parameterized by size. With these primitive state elements, we can define RAMs, ROMs, register files, and other useful components. However, it is important to understand that operations as defined here always occur within a single cycle, and thus for analysis, all state elements are constrained to appear as sources or sinks.

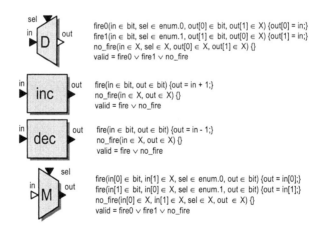

Figure 4.3.   Actor semantics.

After defining the actors, an architect composes them by connecting their ports with signals in a hierarchical schematic editor. Most architects will use predefined library actors which are parameterized by port width. An example composition is shown in Figure 4.4. We purposely leave ports unconnected if we want these ports to be programmable. The non-determinism of these ports will be leveraged in the operation view to generate the supported operations of the micro-architecture.

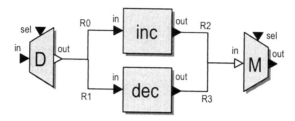

Figure 4.4. Example architecture.

We have developed a particular architecture view based on a constraint-based functional language, but other architecture views that are capable of describing our wide range of architectures could also be used or developed. For example, the MIMOLA ADL [150] could be used. In fact, our architecture view is similar to MIMOLA except that we abstract control values. The abstraction enables delaying the encoding of control until it is actually needed in the hardware view. Furthermore, we automatically generate the encoding.

3.1.4   Operation View.   The operation view provides the designer the ability to extract all interesting configurations of the datapath described in the architecture view. To extract the operations, we first translate the design to first order logic. Figure 4.5 demonstrates the formulation of the architecture shown in Figure 4.4. The possible colors for each signal are defined as a set. Each actor has a set (named *actor.fire*) that encodes which of its firing rules are active. Because the validity constraint for each of these actors is a simple disjunction of its rules, the set is constrained to be non-empty.

After formulating the architecture as a first order logic expression, we then find the set of satisfying solutions. These solutions represent the operations supported by the architecture. The solutions are found using an iterative SAT procedure shown in Figure 4.6 called *FindMinimalOperations* (FMO). The procedure must be restricted to find "minimal" solutions, as there are generally an exponential number of solutions, scaling with the amount of independent parallelism in the design. A solution is "minimal" when no additional signals can be "X" (or not present) and still satisfy the model. The inner loop of FMO does this minimization. Following the creation of an operation, we restrict the model formula so that subsequent "minimal" operations are not simply combinations of previous operations. By limiting FMO in this manner, we can quickly find the supported operations. Either the assembly or compiler view

D.in $\in$ {X, χ}, D.sel $\in$ {X, 0, 1}, R0 $\in$ {X, δ}, R1 $\in$ {X, ϵ}, R2 $\in$ {X, ϕ}, R3 $\in$
{X, γ}, M.sel $\in$ {X, 0, 1}, M.out $\in$ {X, α, β}, D.fire $\subseteq$ {fire0, fire1, no_fire},
inc.fire $\subseteq$ {fire, no_fire}, dec.fire $\subseteq$ {fire, no_fire}, M.fire $\subseteq$ {fire0, fire1, no_fire}

((D.in = χ $\wedge$ D.sel = 0 $\wedge$ R0 = δ $\wedge$ R1 = X) $\Leftrightarrow$ (fire0 $\in$ D.fire)) $\wedge$
((D.in = χ $\wedge$ D.sel = 1 $\wedge$ R0 = X $\wedge$ R1 = ϵ) $\Leftrightarrow$ (fire1 $\in$ D.fire)) $\wedge$
((D.in = X $\wedge$ D.sel = X $\wedge$ R0 = X $\wedge$ R1 = X) $\Leftrightarrow$ (no_fire $\in$ D.fire)) $\wedge$
(D.fire $\neq$ $\varnothing$) $\wedge$

((R0 = δ $\wedge$ R2 = ϕ) $\Leftrightarrow$ (fire $\in$ inc.fire)) $\wedge$
((R0 = X $\wedge$ R2 = X) $\Leftrightarrow$ (no_fire $\in$ inc.fire)) $\wedge$
(inc.fire $\neq$ $\varnothing$) $\wedge$

((R1 = ϵ $\wedge$ R3 = γ) $\Leftrightarrow$ (fire $\in$ dec.fire)) $\wedge$
((R1 = X $\wedge$ R3 = X) $\Leftrightarrow$ (no_fire $\in$ dec.fire)) $\wedge$
(dec.fire $\neq$ $\varnothing$) $\wedge$

((R2 = ϕ $\wedge$ R3 = X $\wedge$ M.sel = 0 $\wedge$ M.out = α) $\Leftrightarrow$ (fire0 $\in$ M.fire)) $\wedge$
((R2 = X $\wedge$ R3 = γ $\wedge$ M.sel = 1 $\wedge$ M.out = β) $\Leftrightarrow$ (fire1 $\in$ M.fire)) $\wedge$
((R2 = X $\wedge$ R3 = X $\wedge$ M.sel = X $\wedge$ M.out = X) $\Leftrightarrow$ (no_fire $\in$ M.fire)) $\wedge$
(M.fire $\neq$ $\varnothing$)

[[δ]] = [[χ]] ; [[ϵ]] = [[χ]] ; [[ϕ]] = [[δ]] + 1; [[γ]] = [[ϵ]] − 1; [[α]] = [[ϕ]] ; [[β]] = [[γ]].

Figure 4.5.   First order logic formulation of example.

then combines these operations in time and/or space to create what would be considered more traditional instructions.

Figure 4.7 shows the resulting solutions found by running FMO on the example architecture in Figure 4.4. Each solution indicates the colors of the signals. The first solution represents the operation to do nothing, the second solution performs the increment operation, and the third solution is the decrement operation. The associated axiomatic relations will allow us to construct logic that computes the values of any sink signal's symbolic colors in terms of the source signals' symbolic colors.

It is common that some of the operations generated are unwanted or have spatial and/or temporal constraints between each other. A spatial constraint indicates that a set of operations always occur together on a cycle, whereas, a temporal constraint specifies a particular order for a set of operations. Using these constraints, we can build traditional instructions by chaining operations. For example, a three-operand register-to-register "add" instruction for a 5-stage DLX would be implemented by adding a temporal constraint to the set of operation constraints. The constraint would indicate that the operations that constitute the "add" in the IF, ID, EX, MEM, and WB stages occur one cycle after another in that pipeline order. Removing operations and constraining relationships between operations allows a hardware view to synthesize more efficient control. For this reason, we view restricting and constraining operations as an integral part of the design process. The unrestricted operations and constraints between these operations represent the programmability of the architecture.

```
BASE              CNF formulation of the constraints of the model
PVAR              set of primary variables
ADDED_PVAR        set of PVARs that are added (each is associated with an element of PVAR)
CERT              satisfying certificate (i.e. set of literals satisfying the BASE)
isPresent         (PVAR x CERT) → Boolean (true if port present in cert)
present           (PVAR) → returns the variable indicating that PVAR is present
isSatisfiable     BASE → Boolean (true if CNF is satisfiable)
getCertificate    BASE → CERT (last certificate that made isSatisfiable TRUE)
getVar            (ADDED_PVAR) → PVAR (returns the associated PVAR)
getVar            (PVAR) → ADDED_PVAR (returns the associated ADDED_PVAR)
```

$$
\begin{aligned}
&findMinimalOperations(\text{BASE}) \{ \\
&\quad \text{OPERATIONS} = \{\} && (1) \\
&\quad \textbf{while } (isSatisfiable(\text{BASE})) \{ && (2) \\
&\quad\quad \textbf{do } \{ \\
&\quad\quad\quad C = getCertificate(\text{BASE}) && (3) \\
&\quad\quad\quad \text{"remove the constraints added in (5)"} && (4) \\
&\quad\quad\quad \text{BASE} \wedge= \Sigma \; \{\neg present(\text{pvar}) \mid (\text{pvar} \in \text{PVAR} \wedge isPresent(\text{pvar}, C)\} && (5) \\
&\quad\quad\quad \text{BASE} \wedge= \prod \{\neg present(\text{pvar}) \mid (\text{pvar} \in \text{PVAR} \wedge \neg isPresent(\text{pvar}, C))\} && (6) \\
&\quad\quad \} \textbf{ while } (isSatisfiable(\text{BASE})) && (7) \\
&\quad\quad \text{"remove the constraints added in (6)"} && (8) \\
&\quad\quad \text{op} \leftarrow createOperation(C) && (9) \\
&\quad\quad \textbf{if } (\text{"first solution"}) \{ && (10) \\
&\quad\quad\quad \text{BASE} \wedge= \Sigma \; \{\text{pvar'} \mid \text{pvar'} \in \text{ADDED\_PVAR}\} && (11) \\
&\quad\quad\quad \text{BASE} \wedge= \prod \{(\text{pvar'} \Rightarrow getVar(\text{pvar'})) \mid \text{pvar'} \in \text{ADDED\_PVAR}\} && (12) \\
&\quad\quad \} \\
&\quad\quad \text{OPERATIONS} \cup= \text{op} && (13) \\
&\quad\quad \text{BASE} \wedge= \text{op} \Leftrightarrow \prod \{present(\text{pvar}) \mid (\text{pvar} \in \text{PVAR} \wedge present(\text{pvar}, C))\} && (14) \\
&\quad\quad \text{BASE} \wedge= \prod \{(\text{op} \Rightarrow \neg getVar(\text{pvar'})) \mid (\text{pvar'} \in \text{ADDED\_PVAR} && (15) \\
&\quad\quad\quad\quad \wedge present(getVar(\text{pvar'}), C))\} \\
&\quad \} \\
&\quad \text{"remove the constraints added in (11), (12) and (15)"} && (16) \\
&\quad \textbf{return } \text{OPERATIONS} && (17) \\
&\}
\end{aligned}
$$

Figure 4.6.   FindMinimalOperations (FMO).

We have developed a particular operation view that utilizes SAT to extract the operations. The MIMOLA effort also provides a method to extract operations using BDDs [148]. However, our approach offers the designer the freedom to delay the control implementation as mentioned in Section 3.1.3. To support this abstraction, we perform a simple transformation of the design to first-order logic. We then apply SAT to extract the operations. In practice, we have found that this formulation is straightforward and easy to solve.

3.1.5   Simulation View.   After describing the architecture and extracting the supported operations, we must provide a means to simulate the operations. Our simulator view provides the capability to generate a high-speed cycle-accurate simulator for the architecture. Given an operation, we translate any necessary axiomatic relations into operations of a language such as C/C++. For each operation found with the static analysis performed in the operation view, we are able to remove all of the control signals or any others which are

NOP: D.sel = X ∧ D.in = X ∧ R0 = X ∧ R1 = X ∧ R2 = X ∧ R3 = X ∧
M.sel = X ∧ M.out = X ∧ D.fire = no_fire ∧ inc.fire = no_fire ∧
dec.fire = no_fire ∧ M.fire = no_fire

inc: D.sel = 0 ∧ D.in = χ ∧ R0 = δ ∧ R1 = X ∧ R2 = φ ∧ R3 = X ∧
M.sel = 0 ∧ M.out = α ∧ D.fire = fire0 ∧ inc.fire = fire ∧ dec.fire = no_fire ∧ M.fire = fire0

dec: D.sel = 1 ∧ D.in = χ ∧ R0 = X ∧ R1 = ε ∧ R2 = X ∧ R3 = γ ∧
M.sel = 1 ∧ M.out = β ∧ D.fire = fire1 ∧ inc.fire = no_fire ∧ dec.fire = fire ∧
M.fire = fire1

Figure 4.7. Satisfying certificates for example.

constant leaving only the relevant state-to-state combinational logic for each instruction (an *instruction* is defined as a set of operations to be executed within a cycle). Furthermore, since the program is specified as a series of symbolic operations, there is no need to simulate or specify any particular encoding of the operations or decoder. The resulting simulator is a compiled-code simulator for any realization of this architecture. An interpretive simulator can be useful if the program to be executed is not known. However, the interpretive simulator suffers in performance compared to the compiled-code simulator.

Our generated simulators are comparable to other top performing simulators such as LISA [186] and JACOB [149]. However, we feel that our simulators have several advantages: They support a wider range of architectures, generating the simulator from the extracted operations is simpler than the LISA approach, and the strong semantics of our underlying model allows for fairly aggressive optimizations and transformations [240]. A more detailed description of the underlying simulation techniques applied by Tipi are described in Section 3.2.

3.1.6 Assembly view. Our assembly view represents the first step towards the generation of a compiler. Given a set of operations, we generate an assembler for that set. The primary task of the assembler is to read an assembly code file and produce the appropriate pre-decoded machine code for our simulator. The assembly code file consists of a list of groups of concurrent operations that are, in turn, to be executed in sequential order. During the machine code generation, the assembler verifies that a group of operations can be "issued" together. To verify this property, we perform a satisfiability check on the model with the union of all firing rules in the group of operations forced to be active. If it is satisfiable, then the group of operations can be issued together. We also check that the spatial and temporal constraints are met. This is important because the simulator (which has the control logic simplified away) does not perform any such check, and clearly we do not wish to simulate operation combinations that are not possible on the actual hardware.

3.1.7   Hardware View.   We can simulate and program our architecture without explicitly specifying the control logic, but to actually realize an implementation of the architecture, we must commit to a control strategy. There are many control strategies. For very simple programs, a dedicated FSM that represents the program can be designed. Another simple control strategy that we have implemented as the default is horizontal microcode. More elaborate control strategies such as reconfigurable control and RISC or VLIW encoding schemes could also be designed. Each control strategy uses the information about the operations supported and the spatial and control constraints to optimize the generated controller. Once a control strategy has been chosen, the hardware view can then produce a synthesizable RTL hardware model by combining the information in the architecture view with the control strategy to realize an implementation. A similar approach has been used with the LISA ADL to automatically synthesize instruction encodings [176]. We then leverage a traditional HDL flow to generate hardware and provide combinational timing feedback for each operation.

3.2   Generating Fast Bit-True Cycle-Accurate Simulators for Programmable Architectures

Architecture description languages have successfully been used to model programmable architectures when the instruction set and corresponding opcodes are known. However, for many application-specific processor starts, designers are primarily concerned about designing the datapath, not the control and instruction set. As shown in the preceding section, Tipi uses a structural description that does not require the specification of the control or the instruction set. Instead, we abstract the architecture as a set of automatically extracted primitive operations. Hardware descriptions and fast bit-true cycle-accurate simulators are then automatically generated from the operations, which is the focus of this section. Results show that our simulators are up to an order of magnitude faster than any other reported simulators of this type and two orders of magnitude faster than equivalent Verilog® simulators. Providing a correct-by-construction and automated path to the generation of simulators and hardware descriptions is an essential element towards a comprehensive exploration of a large design space.

3.2.1   Introduction.   Currently, architecture description languages (ADLs) are being used to design and simulate application-specific instruction processors (ASIPs). Since ADLs require the specification of the instruction set and binary opcodes, they are especially useful when such things are known. However, as designers are creating new complex and irregular application-specific programmable architectures, they may not know the exact details of the instruction set they want to support and/or the binary encoding of the opcodes. Moreover, defining the instruction set for their architecture may be complicated by irregularities and complexities in the underlying datapath.

In fact, we believe that ASIP designers are more concerned with finding the correct mix of functional units, memories, and required routing in order to meet the requirements of an application domain than they are with defining the instruction set. Defining the control and instruction set is certainly an important aspect of the design, but this should be an automated task based on the capabilities of the underlying datapath. Furthermore, manually developing the single-cycle instruction set abstractions found in ADLs is complicated by such things as multiple memories and forwarding paths.

Designers benefit from structural descriptions in Tipi where the control and instruction set can be automatically extracted from the datapath. We will show that this approach is particularly beneficial for the automatic generation of bit-true cycle-accurate simulators. In fact, our results reveal that this approach leads to faster simulators than ADL-based frameworks.

In order to accommodate designers developing new ASIPs, we have developed a design framework with the following key features:

- The specification of the architecture structure is allowed.

- A large, expandable library of components is included.

- Control and instructions do not have to be specified.

- Operations (instructions) are automatically extracted.

- Control and opcodes are automatically generated.

- Synthesizable RTL Verilog® can be generated.

- Bit-true cycle-accurate simulators can be generated.

- Assemblers can be generated.

By allowing designers to focus solely on the datapath and by providing fast simulators and synthesizable RTL, our framework enables the fast design and implementation of programmable architectures.

This subsection is organized as follows. In the next subsection, we provide a brief overview of our language and the extraction of primitive operations. In Subsection 3.2.3, we discuss the generation of simulators, which is the primary focus of this subsection. In Subsection 3.2.4, we elaborate on the generation of hardware. We review related work in Subsection 3.2.5. Simulation performance results are part of our case studies in Section 4.

3.2.2 Tipi Design Framework. We have developed the correct-by-construction framework Tipi for designing architectures. The core of this framework is a constraint-based functional language [239]. We can describe any single-clock synchronous system in our framework. Furthermore, we can extract the primitive operations of the architecture as described in the preceding section. From the primitive operations, we automatically generate bit-true simulators and a synthesizable RTL description of the architecture. The complete flow of the design methodology is shown in Figure 4.8. In this subsection, we will focus on the generation of simulators and synthesizable RTL, which is shaded in the figure.

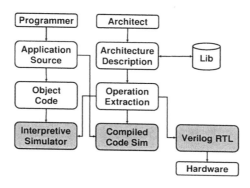

Figure 4.8. Tipi design methodology.

Description of Architectures. In our framework, architectures are composed in a hierarchical schematic editor by connecting ports on components with relations. Each component is described in a constraint-based functional language. Components are described in terms of first-order logic on a set of rules. Rules have an activation and an action section. The activation section indicates what signals have to be present and/or not present in order for the rule to be activated. The action section indicates what operations to perform if the rule is activated. An add component is shown below as an example.

```
add (input inA, input inB, output out) {
  rule fire(out, inA, inB)
    { out = inA + inB; }
  rule no_fire(-out, -inA, -inB)
    { }
  (fire || no_fire);
}
```

The component is interpreted as follows. The adder is valid if and only if either the rule *fire* inclusive-or the rule *no_fire* are activated. The rule *fire* is activated when *inA*, *inB*, and *out* are present. When *fire* is activated, the operation *out* = *inA* + *inB* is performed. The rule *no_fire* is activated if *inA*, *inB*, and *out* are not present. All other combinations of *inA*, *inB*, and *out* are considered invalid and would be reported as an error. Such a formulation will be leveraged to extract the primitive operations of the architecture.

In order to support reuse and ease of use, the component language utilizes a number of key features shown below.

- The language supports the operators, constant declarations, and type system found in Verilog®.

- Components can contain any number of state elements.

- The language is type-polymorphic based on a Boolean type and N-bit unsigned integers.

- The types of ports and intermediate results can be specified as expressions on constants and/or other types found in the component.

- Ports on components can be treated as lists in order to support components that can work on a variable number of ports.

These features allow designers to easily construct highly parameterized components for reuse. The following N:1 mux component demonstrates the utility of a number of these components.

```
mux (output out, input ⟨out.type⟩ @din, input {enum} sel) {
  foreach(i) {
    rule fire(out, din[$i], foreach(j) if ($i != $j) -din[$j] , sel)
      { out = din[$i]; }
  }
  rule no_fire()
      { }
  or(fire) || no_fire;
}
```

The syntax @*din* indicates that *din* is a list that must take on the type of *out*. The syntax {*enum*} indicates that signal *sel* is an enumerated control signal where *sel* takes a new value for each rule activation list that it is included in. The *foreach* and indexing syntax is used to create the appropriate number of rules given the width of the widest port, in this case, *din*. If *out* had type ⟨10⟩ and there were two relations connected to *din*, then the expansion of the 2:1 mux would be as shown below.

```
mux (output ⟨10⟩ out, input ⟨10⟩ din[0], input ⟨10⟩ din[1], input ⟨0⟩ {enum} sel) {
    fire_0(out, din[0], -din[1], sel)
        { out = din[0]; }
    fire_1(out, -din[0], din[1], sel)
        { out = din[1]; }
    no_fire (-out, -din[0], -din[1], -sel)
        { }
    (fire_0 || fire_1 || no_fire);
}
```

The N:1 mux example also demonstrates the fact that one does not need to specify control. The signal *sel* is not used in the action section of any of the rules. The actual value of *sel* is only needed for hardware generation and can be abstracted during the generation of the simulators. Alleviating the need to specify control allows designers to focus their attention on designing datapaths. We have found that this coupled with a library of reusable components greatly increases the productivity of designers.

Extraction of Operations.   After specifying an architecture as a network of components, we can then extract the primitive operations that the datapath supports. An operation is a combinational transformation between sources (input ports or state reads) to sinks (output ports or state writes). A primitive operation is an operation that is not simply a combination of other operations and does not contain any simpler operations. Extraction is performed by creating a constraint expression from the network of components using their rule constraints. All minimal solutions which correspond exactly to minimum operations are then found for the constraint expression. The specific details of extracting primitive operations have been described in the preceding section. After extracting the primitive operations, we can now generate bit-true cycle-accurate simulators as well as synthesizable RTL.

3.2.3   Simulator Generation.   The basic structure for our generated simulators is straightforward. For each cycle, we run a statically determined instruction. An instruction is defined as an unordered set of operations. Each operation may contain a set of statically determined parameters and may read inputs from and write outputs to the interface of the architecture. Decoding an operation simply requires jumping to an appropriate label based on the name of the operation. For each operation label, we execute the appropriate source to sink transformations. Since multiple operations can be executed in a single-cycle, we commit writes to state elements at the end of a cycle. The resulting simulator is equivalent to a discrete-event simulation of an FSM in an RTL simulator. However, our simulator is much faster because we can statically determine the schedule.

The generated simulators are implemented in C++ for performance reasons. In order to get the utmost performance from our simulators, we use a number of constructs that would not be found in hand-written code. For example, we use computed gotos, template meta-programming, and inlining. As we will see later, these techniques coupled with a good compiler result in high performance simulators.

Handling Arbitrary Bit-Widths.   As mentioned previously, we provide support for arbitrary bit-widths. At a base, we support a templated data type called UINT⟨width⟩. Based on the value of width, the appropriate underlying type is created as is shown next.

width	type	defined
1-32	uint32_t	stdint.h
33-64	uint64_t	stdint.h
⟩ 64	mpz_class	gmpxx.h

The standard library "stdint.h" is used for widths less than or equal to 64. For widths greater than 64, we leverage the GNUmp libraries[4]. We choose the appropriate type and libraries in order to get the best performance possible. Furthermore, since the standard C++ operators are defined for each of these types, generating code for expressions is simplified by operator overloading.

However, we do have to define a couple of functions to deal with ranges and residual effects of computing with a fixed-width host architecture. These functions are implemented using template meta-programming in order to make them as efficient as possible. The functions are shown next.

```
range⟨msb,lsb,width⟩(f)
assign⟨msb,lsb,width⟩(f, g)
crop⟨width⟩(f)
```

The *range* function extracts the bits between *msb* and *lsb* in *f* (*f* has width *width*). If *msb* is less than *lsb*, then the bits are extracted and reversed. The *assign* function assigns *g* (*g* has width *width*) to *f* between the specified *msb* and *lsb*. Again if *msb* is less than *lsb*, then the bits in the range are reversed. The *crop* function is used to mask residual bits that exist in a *uint32_t* or *uint64_t* type. This is required when accessing memories because other operations may

---

[4]*http://www.swox.com/gmp*

have created residual bits in the upper portion of the 32 or 64 bit type. If no memory access occurs, then these residual bits can be ignored.

The definition of UINT, *range*, *assign*, and *crop* makes code generation straightforward. For example, if we wanted to add a 67 bit number, *a*, and 34 bit number, *b*, and then write bits 7 through 3 of the resulting sum to the upper 5 bits of a 10 bit number, *c*, then we would generate the following code:

```
UINT⟨67⟩::T a = value₁;
UINT⟨34⟩::T b = value₂;
UINT⟨10⟩::T c = value₃;
assign⟨9,5,5⟩(range⟨7,3,67⟩(a + b), c);
```

Although this approach works well for types with 64 or less bits, performance degradation is witnessed for types above 64 bits. Besides having to perform non-native arithmetic for *mpz_class* types, the compiler does not perform the appropriate management of temporaries. In the future, we plan on manually managing temporaries in the simulator.

Interpretive Simulator. Generating an interpretive simulator is useful when the designer does not want to create and compile a new simulator for each program. The interpretive simulator can read and interpret object files created by the automatically generated assembler. The basic structure of the generated interpretive simulators is shown below.

```
while(!testbench→exit()) {
  instruction = testbench→getInstruction();
  foreach(operation ∈ instruction) {
    switch(operation) {
      case op_1:                    // perform operation 1
      case op_2:                    // perform operation 2

      case op_n:                    // perform operation n
    }
  }
  // commit writes
}
```

Along with the simulator, a testbench is automatically generated. The testbench advances time, provides inputs, and accepts outputs from the simulator. If the testbench does not stop the simulation, then a new instruction is read and all of its operations are simulated. The assembler guarantees that there are no conflicts between operations in an instruction. Therefore, the execution order of the operations within an instruction does not matter. After simulating all the operations, the writes are committed and the cycle count is increased.

The control of the simulation is orchestrated through a special port called "instruction" in the architecture. At the beginning of each cycle, the last value written on "instruction" is interpreted as the instruction to execute. Before beginning the simulation, the object file is loaded into a special component called "InstructionState". Connecting the output of "InstructionState" to the "instruction" port would result in an architecture that generates its next instruction. Although this is the most common control strategy, it is not required.

For each operation, the actual expressions to be performed are automatically extracted from the action sections of the components which are activated for a given operation. We apply copy propagation on the network of activated expressions for a given operation. If we did not do this, we would have a number of unnecessary temporaries resulting from the component netlist. Dead code elimination is also applied to improve the quality of the generated code. Although compilers will attempt to perform these optimizations, we have found that making these optimizations before compilation is beneficial.

After optimizing the expressions of an operation we emit the operation section as shown below:

```
case op_i :
          // read operation parameters
          // read inputs
          // execute expressions
          // write outputs
```

First, the operation parameters and required external inputs are read. Next, the set of expressions that use these parameters, inputs, and memories are executed and any pending writes are indicated (the writes are committed at the end of the cycle). Finally, any writes to external outputs are made.

Compiled-Code Simulator.   Compiled-code techniques can be utilized to further improve the performance of the simulator. If we know the program that is going to be run, we can hard code the operations into the simulator. The basic structure of the generated compiled-code simulators is shown below.

```
_0:    testbench→tick()
          // perform instruction 0
          // perform computed goto

_n:    testbench→tick()
          // perform instruction n
          // perform computed goto
```

Before runtime, we know what operations are included in each instruction. Therefore, we can combine the operations to create a single set of expressions for each cycle. We then apply the same optimizations as we did for the interpretive simulator. Combining the operations of an instruction removes the overhead of iterating through the list of operators in an instruction. Furthermore, we utilize computed gotos to jump between runtime computed labels. This removes the need to read the next instruction. A further optimization that we will make in the future is to remove gotos when we can statically determine that the simulation simply proceeds to the next block. The resulting simulator simply jumps between labels to simulate the program until the simulation exits. The call to *testbench→tick()* increments the cycle count and determines if the simulation should stop.

Probe Components. In order to analyze and debug the simulation, we allow the insertion of probes in the network. Components that utilize our probe syntax are required to create constraints that do not add any new semantics to the design. This guarantees that they can be safely removed before creating the implementation. The probe syntax provides a means to capture the trace of the simulation.

Black-Box Components. We provide support for specifying a component directly in C++. This is useful when the component does not need to be synthesized or an implementation for the component already exists that is verified to be consistent (i.e. IP integration). For example, we may want to write a complex routine to analyze the behavior of traffic to and from a cache. Specifying such a component in our language is unnecessary since this component would not be included in the implementation. For certain memory components, such as a CAM, we would write a black-box component. We would not want to synthesize a CAM, but would instead use a predefined CAM block for the implementation.

Testbench. As mentioned previously, a testbench is also generated for each simulator. The testbench creates a function for each input and output port on the interface. Furthermore, a control function exists that is called at the beginning of a cycle to control the advancement of time and to determine if the simulation should stop. The testbench functions can be customized to provide an environment for the simulation. Furthermore, a system simulator (e.g. SystemC™ based) could provide the inputs, accept the outputs, and implement the control function in a manner that creates a master-slave relationship between the simulators.

3.2.4 Hardware Generation.

A key component of our design methodology is to be able to produce synthesizable RTL from our architecture descriptions. Since our components are specified in a structural manner using syntax and type rules consistent with Verilog®, it is straightforward to generate the appropriate combinational and sequential hardware for the design. However, unlike our simulators, we cannot implement each operation as a hardware path. Instead we preserve the structure of the architecture, and implement a controller that multiplexes the paths in the architecture appropriately.

For each operation, we can determine what control signals and write enables are required in order to enable the appropriate paths in the architecture. We then use this information to create a horizontally micro-coded controller. In order to simulate a program, we also generate the appropriate control words to be embedded in the program memory. Currently, we are exploring an encoder-decoder scheme that compresses the instruction stream based on the analysis of static program traces. Although it is beyond the scope of this subsection, the basic idea is presented here, which is to compress the micro-code in software and then decompress it in hardware to get the appropriate control.

> compiler(SW) → encoder(SW) →
> program store(HW) → decoder(HW) →
> micro-code buffer(HW) → micro-code control(HW)

Our current scheme simply implements the encoder and decoder as the identity function. The benefit of this general approach is that the datapath and control do not change when the encoder and decoder change.

3.2.5   Related Work on Simulation.
A number of approaches to retargetable simulation based on ADLs have been proposed. Frameworks, such as FACILE [213], ISDL [92], Sim-nML [94], are optimized for particular architectural families and cannot capture the range of architecture that we can. More flexible modeling that supports both interpretive and compiled-code simulation is presented in the LISA [175] and EXPRESSION [198] frameworks. All of these approaches require that the designer specify the instruction set, and thus are more suitable for modeling architectures where the ISA is known. Although we have not applied the just-in-time techniques presented in the most recent work, we have applied a number of the optimizations including compiled-code techniques, static analysis, and compiler optimizations.

The MIMOLA framework most closely resembles our approach to retargetable simulation. Interpretive and compiled-code simulators have been generated from structural MIMOLA descriptions [149]. Since the structure of the datapath and control are specified in MIMOLA, hardware generation is also straightforward. The key difference between our approach and MIMOLA lies in the fact that we do not require the control encoding to be specified, thus allowing the designer to focus on the design of the datapath. We believe that our more aggressive simulator generation optimizations have allowed us to outperform the JACOB simulators. Finally, we have detailed a different method for handling arbitrary bit-width data types.

4. Tipi Case Study

The following case study illustrates the Tipi design flow – correct by construction, multi-view, operation-level design – applied to two examples: an application-specific coprocessor and a general purpose computation core.

4.1 Design Entry

The following examples of an application-specific and a general purpose processor show the effectiveness of Tipi's approach to describe datapaths and their instruction set.

4.1.1 Channel Encoding Processor. To demonstrate and evaluate our tools and methodology, we supported an experienced industrial ASIC designer who implemented a channel encoding processor using our methodology (Figure 4.9). The processor is capable of performing CRC, UMTS Turbo/convolutional encoding, and 802.11a convolutional encoding. The design is composed of approximately 60 actors which are divided into a "control" plane composed of a PC and zero-overhead looping logic, and a "data" plane composed of a register file, accumulator, and bit manipulation unit. The division of planes is for the designer, not the tool. The designer chose to leverage our default control strategy generation (horizontal micro-code) by leaving the select ports of the multiplexers, the register file address ports, and immediate value ports unconnected. The non-determinism of these unknown values represents the programmability of the architecture; the generated controller must produce appropriate values for these ports on each cycle.

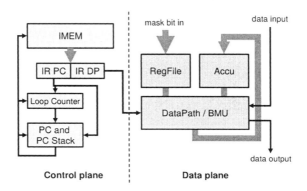

Figure 4.9. Channel Encoding Processor (CEP).

While designing the processor, we were only focused on creating a pipeline that ran applications in our domain well. We were not concerned with the instruction set. A screenshot of the design in our development environment is shown in Figure 4.10. The screenshot shows both the schematic editor and the extracted operations. The diagram only shows the top level of the design (many of the components are hierarchical). A total of 46 operations were extracted from the design, of which half were unwanted and thus restricted.

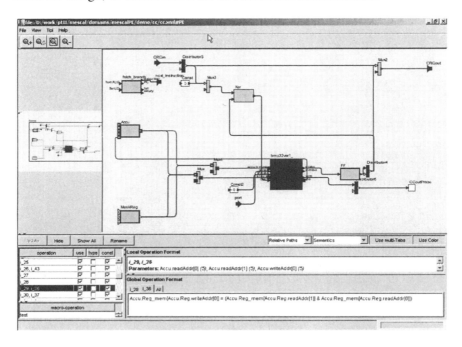

Figure 4.10.   Channel Encoding Processor (CEP) screenshot.

To program the processor, each instruction generally consists of "control" and "data" operations. The "data" operations are specialized bit manipulation operations for CRC and convolution encoding. Due to resource sharing, most of these operations are mutually exclusive. The "control" operations are things such as increment PC, set loop counter, decrement loop counter, and test end of loop. These can be combined to form an instruction that, for example, decrements the loop counter, branches if the loop counter is zero, otherwise increments the PC. The instruction also includes additional data plane operations. The applications targeted for the processor contain many streaming tasks that require multiple cycles per sample. To implement these applications, the designer created zero-overhead loops containing small sets of instructions to perform each task. Effectively a designer can use an ASIC methodology to design a programmable architecture by specifying the requisite "control" and "data" planes and leverage our methodology to extract the operations and generate the simulator and assembler.

4.1.2 DLX Model.   Although our framework is mainly targeted towards application-specific programmable cores, in order to compare the effectiveness of our approach for designing programmable micro-architectures with existing methods, we have modeled a general-purpose processing core. That means, our model incorporates the characteristic elements of the micro-architecture of the DLX micro-processor [182]. However, since we extract the set of supported operations automatically from the description of the datapath, we do not anticipate matching the binary encoding. Furthermore, the modeled 32-bit DLX is a horizontally micro-coded core supporting arithmetical and logical operations with a five stage pipeline, see Figure 4.11. Conditional jumps are supported. We extracted 113 primitive operations from the datapath that we combined to provide 30 pipelined assembler instructions. The model includes instruction and data memory, as well as program counter logic.

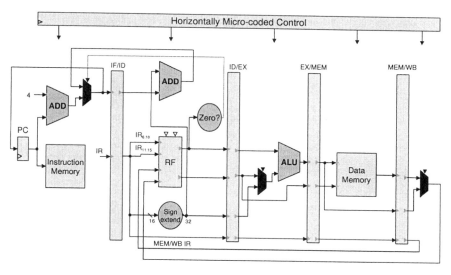

Figure 4.11.   DLX microarchitecture as modeled in Tipi.

We have implemented three representative benchmark kernels in assembly code: Cyclic Redundancy Check (CRC), Inverse Discrete Cosine Transform (IDCT), and a signal processing multiply and accumulate filter loop including masking, as used by established benchmarks (EEMBC, DSPStone, and Mediabench). The lookup-based 32-bit CRC requires only a few arithmetic operations, but relatively frequent memory accesses, whereas the complex IDCT has more arithmetic and program flow constructs, but fewer memory accesses. The FIR filter loop in particular allows us to stress the pipeline. As a corner case, we also simulate executing NOP operations only.

The combined runtimes of each view are low enough to enable design space exploration. All experiments were performed on a 1.33 GHz Athlon^{TM} with 768MB of memory. For the channel encoding processor, it takes 24 seconds to generate the operations, five seconds to generate the interpretive or compiled-code simulator (cycle-accurate, bit-true), and half a second to compile the simulator. Our views are all written in Java, but we use a SAT solver written in C for the FMO algorithm. In order to get the highest performance simulator, we generate C++ and use gcc3.2 with -O3 to create the simulator. Each operation requires on the order of ten host machine instructions for decoding control (which is only present in the interpretive simulator), and roughly one instruction for each integer computation. These performance numbers are on par with existing pipeline and cycle accurate simulators.

4.2 Simulator Generation

We designed the architectures in Tipi on our IDE, and then automatically extracted the operations from which we generated interpretive and compiled-code simulators. We also generated synthesizable RTL Verilog®. Since our C++ simulators are equivalent to RTL simulation, we compare our simulation with Cadence's NC-Verilog® simulation of our generated RTL Verilog®. Our simulators were compiled using gcc v3.2 with -O3 and were run on a 2.4 GHz Pentium 4® with 1 GB RAM. NC-Verilog® ran on a dual 900 MHz 64-bit UltraSparc® III with 2GB of memory. In order to compare the results, we liberally scaled the NC-Verilog® numbers by a factor of 2.67 (2400 MHz/900MHz).

4.2.1 DLX Model. The achieved simulation speed results for the DLX are listed in Table 4.2.

Table 4.2. DLX simulation speed results.

design	NC-Verilog®	interpretive	compiled	ops/inst
NOP	2.5 MHz	6.9 MHz	588 MHz	1
CRC	455.7 KHz	4.6 MHz	85.5 MHz	9.12
IDCT	444.5 KHz	4.6 MHz	49.0 MHz	8.65
FIR	361.7 KHz	4.0 MHz	40.8 MHz	5.5

The results are reported for 2 billion simulated cycles. We report the virtual running speed on the host in cycles per second, and the ratio of the average number of primitive operations to equivalent DLX pipelined instructions. For instance, an add instruction needs ten primitive operations to execute on our five stage pipeline.

4.2.2   Channel Encoding Processor.   We also developed a channel encoding processor capable of performing convolutional encoding for applications such as UMTS Turbo encoding. This architecture is more representative of the types of designs our tool is targeting.

For this design, we experimented with various bit widths for the datapath. We only had to write the convolution encoding program in assembly once, since our tool automatically adjusts for bit-width changes if possible. The results of the experiment are shown in Table 4.3.

Table 4.3.   CEP simulation speed results.

bit width	NC-Verilog®	interpretive	compiled
2	533.6 KHz	5.5 MHz	169.5 MHz
4	512.5 KHz	5.7 MHz	169.5 MHz
7	516.5 KHz	5.7 MHz	158.7 MHz
10	507.2 KHz	5.5 MHz	157.5 MHz
13	505.3 KHz	5.4 MHz	157.5 MHz
16	505.3 KHz	5.4 MHz	157.5 MHz
32	491.2 KHz	5.5 MHz	168.1 MHz
33	485.6 KHz	5.2 MHz	132.5 MHz
64	468.8 KHz	5.5 MHz	131.6 MHz

The results are reported for 2 billion simulated cycles. We report raw speeds since the ratio of instructions to operations is not as relevant with this type of processor. The processor was running on average five primitive operations per cycle. The running times are only slightly effected by the bit width size. The largest effect was seen when our simulators were using bit widths greater than the native 32 bit datapath of the Pentium 4®.

4.3   Discussion

The actual time it takes to create these designs was on the order of minutes to an hour. The extraction of operations for these designs, generation of simulators, and generation of Verilog® was performed in a matter of seconds. The majority of the design effort was focused on specifying the programs. This effort required renaming the operations for debugging purposes, controlling the pipeline on a cycle to cycle basis, and specifying macros to ease programming. We have a compiler that alleviates the need to rename the instructions or create macros by supporting a higher level programming language. The compiler is still being tested so we did not use it for our experiments.

The typical speed of the compiled simulators in number of cycles is about a factor of 20 to 60 slower than the native host speed. The compiled C++ simulator is approximately one order of magnitude faster than the interpretive version and is two orders of magnitude faster than a commercial, highly-optimized Verilog® simulator.

Comparing to results in related work, the speed of our simulators stays ahead or keeps up with any other simulation technique. SimpleScalar, a cycle-accurate interpretive simulator used for micro-architecture research of certain classes of general purpose processors, usually achieves speeds four to five magnitudes slower than the host. SimpleScalar models contain more advanced details like branch prediction logic and the simulation of cache hierarchies. However, the models are bound to a few architecture templates without providing bit-accurate precision. In the domain of ADLs, recent instruction set simulation results have been reported for ARM7™, SPARC®, and VLIW cores [175, 198]. When we scale the reported MIPS results to our simulation host, we recognize that we can keep up with the speed of the interpretive simulators. However, our compiled simulators are at least a factor of two faster than ADL-based compiled simulators (this may be a side-effect of our small kernels). When compared to the MIMOLA-based JACOB simulator [149], which we are most closely related to, we find that our simulators are an order of magnitude faster. Our aggressive simulator code generation techniques coupled with a powerful compiler has enabled us to achieve such high simulation speeds.

5.   Designing Memory Systems with Tipi

In this section, we show how the Tipi design flow can be applied to describe further aspects of a programmable platform, in this case the memory subsystem. The Tipi framework contains a special view for this purpose. Internally, Tipi's techniques for correct-by-construction, multi-view, operation-level design are reapplied so that several memories and programmable cores can efficiently be designed and evaluated concurrently. In this way, whole platforms can be evaluated.

The goal of the memory domain in the Tipi framework is to allow the designer to comprehensively explore the memory space at the architecture level by giving facilities for rapid construction and rapid feedback. The designer can then quickly get an intuition of the interesting areas of the design space and the quality of the final memory design. To do this, a library of basic elements is provided that can be configured and composed to easily model many memory systems. This library can be extended by the designer to model more exotic memory systems. From the memory system models, C-based simulators can be automatically generated through a series of steps to give quick feedback on the performance of the system. In the following sections we first discuss the library of memory elements that is provided. We give an overview of the flow for the automated simulator generation that is based on Tipi's simulator generation described in Section 4.2. Later, we give details on each element of the flow. To conclude, we review previous work on memory system evaluation.

5.1   Memory Design Space Element Library

The Tipi framework can model most memory systems. To speedup the modeling process, a library of pre-built configurable memory elements is provided. The library includes commonly used memory elements, such as RAM, ROM, and cache. Configurability is provided at the architectural level. For example, a cache element is parameterized by associativity, block size, set size, replacement policy, write allocation policy, and write back policy. The library elements are data polymorphic and therefore can work with data types of any bit width. Using the configurability and data polymorphism, these elements can be connected and configured to efficiently model a wide range of memory systems. Figure 4.12 shows a simple example with a cache element and a memory element connected to form a two-level memory hierarchy. The unconnected box in the figure is to indicate the semantics of the memory domain. The exact semantics will be described in a later section.

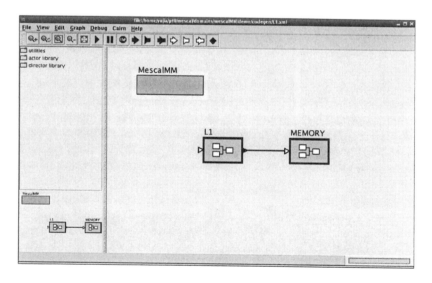

Figure 4.12.   Two-level memory hierarchy in Tipi's memory domain.

5.2   Simulator Generation Flow

From a memory design description created by composing and configuring library elements, a C-based simulator can be automatically generated. The flow for the generation process is summarized in Figure 4.13. Starting from the left is the original high level design described in the memory domain. The memory domain is formalized and has specific semantics. The high level design is first translated into a low level design. The low level design is described in the multi-processing element (PE) domain. This domain includes multiple processing elements described in Tipi that can communicate with each other. The translation process is facilitated by the formalism within both design levels that ensures the correctness of the process.

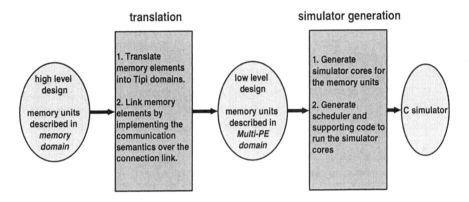

Figure 4.13.   Design flow for the generation of a Tipi memory simulator.

The translation stage itself is composed of two parts. First, each memory element is translated into the Tipi processing element (PE) domain that is defined by the views described in Section 3.1. Then, each translated element is linked by implementing the communication semantics of the corresponding memory domain link. From the low level design, the C-simulator is then generated. The simulator generation stage is another two part process. First, a simulator core for each low level memory unit is generated using the Tipi simulator process described earlier in Section 4.2. Then a scheduler and supporting code is generated to orchestrate the simulation of the cores. The generated scheduler maintains the communication semantics specified in the multi-PE domain and thus coordinates the individual simulators.

If custom memory elements are required, designers can still leverage the flow in two ways. In the first method, the designer has to model a custom element using basic elements of the design library. In the second method, the designer must provide the translation of the element to the low level design explicitly, which is the first part of the translation stage. All following stages of the flow can be reused.

The flow described above generates a C-simulator of the whole system. In order to simulate the memory system, we need some input that drives the memory system. Currently, there are two ways to specify this input. The first method uses a memory access trace that triggers the execution of the memory system simulator. Here, an additional element is added to the high level design that inputs a trace to the memory system. The element sequentially passes the memory requests of the trace to the memory system and waits for the completion of a request before proceeding to the next. While this method is sufficient for evaluating a uniprocessor memory system, it does not work for a multiprocessor memory system. In a multiprocessor scenario, the performance of the memory system can affect the program behavior and therefore the memory access behavior of the application. The trace alone can not model this dependency. To solve this, the second method executes an application on a single or multiple-processor model that is connected to the memory system. Here, additional Tipi elements that model the processors running the application are added in the high level design. The inclusion of Tipi processor models allows the system to model the program behavior changes due to the memory system performance at run-time. For both methods the added element is carried through the entire simulator generation flow and becomes part of the final simulator.

In the following sections we give more details on the semantics of the memory domain and then the multi-PE domain. Later, we discuss the details of the translation stage and the simulator generation stage.

5.3   Memory Domain

The memory domain is used to describe the memory design at a high level, as shown in Figure 4.12. Formalization enables and facilitates many useful features, such as automatic simulator generation, by providing a framework that can precisely and naturally specify the requirements. Within this framework, similar to the architecture view in Tipi described in Section 3, the basic element is an actor. An actor contains ports through which it can communicate with other actors. The actors are categorized into three disjoint types: Table actors, access actors, and arbitration actors. Any memory component is built by configuring and connecting these actors.

- The *table actor* is a storage element with only input ports. It contains m by n entries of arbitrary data. The table actor does not contain logic. It is simply a group of state elements. Its purpose is to show the structure of the memory.

■ The *access actor* is a logic element that accesses other actors to store and retrieve data. It can contain both input and output ports. It performs operations to determine where and how to access the data through actors. The actor accessed can be any of the three actor types. For example, a cache access actor can access a table actor for the local storage and access a RAM access actor for the next memory hierarchy level. If we have a split cache with both data and instruction cache, then the cache access actor will access the main memory indirectly by first going through an arbitration actor.

■ The *arbitration actor* is a local scheduler that places an order on multiple accesses by different actors to a shared actor. It contains both input and output ports. It is placed before the contended actor to filter through all the accesses.

The memory actor communicates with another actor through ports and a relation connecting the ports. Two types of ports can be used depending on the direction of communication. Communication goes through the relation via function calls semantics. An actor executes a function when it receives that function call through an input port. When executing a function, an actor can initiate function calls to other actors. Once the function finishes the actor returns the result through the same port it received the call. The data type for the argument of the function call is polymorphic. Therefore, these actors work with any data bit width. For type checking, each output port contains a list of functions that it can call, and each input port contains a list of functions that it will accept. An output port can connect to an input port only when all its functions are accepted by the input port.

The table actor is configured by specifying the m by n array size. The content of the table actor can also be modified by the designer. The access and arbitration actors are described by specifying each of the functions they support. The functionality specification is currently done in Java due to ease of usage. However, due to its wide scope it hinders the translation stage of the generation process. Currently, for designer defined customized actors, the translation stage requires manual modifications. In the future, Java will be replaced by a more restricted language for the custom actor specification. A library of access and arbitration actors is provided. The memory element library discussed earlier is built on top of this library.

As an example for how these actors can be connected to form a memory element, we illustrate the construction of a simple RAM library element in Figure 4.14. As shown, the RAM element is composed of two actors, a table actor and an RAM access actor. The table actor contains m by one entries, where m corresponds to the size of the RAM. The table actor's input is connected to the RAM access actor's output. The RAM access actor supports both read and write functions. The read function takes an address and returns the data at that address in the table actor. The write function takes an address and data and writes the data in the table actor at the specified address.

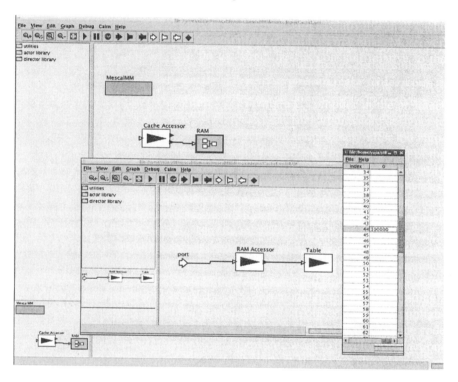

Figure 4.14. RAM model with RAM access and table actors.

5.4   Multi-PE Domain

The multi-PE domain contains multiple Tipi processing elements (PEs). These PEs use the same clock period or a multiple of a base clock period. The communication is carried through the relations connecting the PEs from output to input ports. All inter-PE relations are automatically buffered. Therefore, communication going to another PE is only seen in the next cycle by the receiver. The buffering of the inter-PE communication is enforced to avoid feed-through across multiple PEs, which can result in longer cycle times or a false result. For example, let's suppose a PE executes an operation and sends the result to another PE without an intermediate buffer so that the second PE executes another operation on the received result. Without the buffer, the cycle time has to be extended to ensure the complete execution of both operations across the two PEs.

The multi-PE domain is used to describe the memory design at a low level. It serves as an intermediate level between the high-level description in the memory domain and the actual C-simulator. Here the table actors, access actors, and arbitration actors from the memory domain are described with more hardware related details. These details include datapath and control logic along with possible software as provided by the corresponding Tipi views described in Section 3.

5.5   Memory Domain Translation

The translation stage of the simulator generation flow translates the high-level design described in the memory domain into the low-level design described in the multi-PE domain (see the memory design flow in Figure 4.13). During the translation stage, the memory domain design is viewed as an application being mapped onto the multi-PE domain. There are various valid mapping strategies from the memory domain to the multi-PE domain that maintain the memory domain semantics. The method described here is one possible choice.

As mentioned earlier, the translation stage is composed of two parts. First, the memory elements from the library are translated into the Tipi description for a processing element. Second, the translated elements are connected by implementing the link semantics of the memory domain. We will first discuss the library element translation and then the link implementation. After describing the translation process, we show an example taken through the process.

5.5.1   Memory Element Translation.   In the process of translating memory elements to Tipi architectures, the elements are first decomposed into the basic actors of table actor, access actor, and arbitration actor. Each access and arbitration actor is then mapped onto a Tipi processing element (PE). These PE processors are specifically generated for each actor and contain specific datapath, control, and possibly program code. A table actor is assumed to have only one access actor connected to it. In the translation, the table actor is included in the access actor's PE. In the case where the table actor is connected to multiple access actors or an arbitration actor, one can transform the table actor into a RAM and have the access actors or arbitration actor then connect to the RAM. The PE generation for each actor is provided for the existing library actors. For custom actors specified by the designer, the PE generation step must be supplied explicitly.

The PE generation process for each actor is composed of three stages and follows the Tipi design flow.

- The datapath needed to support the actor's functionality is generated.

- The operations/instructions required by the software implementing the functionality are extracted from the datapath description.

- The software implementing the functionality is generated. These programs implement the arbitration and access actors described in the memory domain.

Datapath and Control Generation. During the datapath generation stage, hardware required by all functions implemented by the corresponding actor is generated. In the simplest case it only includes a datapath. For example, in the case of a stack, the datapath includes logic and storage for push, pop, empty, and full operations. Empty and full operations are used to query if the queue is full or empty. In this datapath, all four operations can execute in a single cycle and simultaneously. No logic for control is required. If the generated datapath shares the logic for several functions or requires multiple executions/iterations to implement the actual function, then control logic is generated to allow a program to control the datapath appropriately. This includes instruction memory, instruction fetch logic, branch logic, and so on. After the datapath and control logic generation, the I/O interface is added. This interface includes the control signals and data signals required for each of the functions listed in each of the ports. The actual signals generated are dependent on the function being implemented and how the link connecting the actors is implemented. For data signals, one signal wire is generated for each argument of the function and one signal wire is generated for each return value of the function. These wires are polymorphic. The actual type resolution for these wires will occur in later stages of the translation process. For the control signals, in the simplest case, a valid wire and a done wire are generated. For both function call initiation and function call returning, the sending side asserts the valid wire to indicate valid data values to send. The receiving side asserts the done wires to acknowledge that data has been received. There are other methods for the control signal generation. The actual method used depends on how the relation in the memory domain is implemented in the linking process.

Software Generation. After the hardware generation, the operation extraction stage is carried out. This stage leverages the Tipi operation-level design methodology explained in the preceding Subsection 3.1 to extract all the required operations from the hardware description. When the control logic and program are not needed as in the case of a stack, only one operation is extracted and the PE generation process stops here. If a program is used to implement the functionality, all operations used by the program are extracted.

If required, software generation takes place after the hardware generation and operation extraction. The core code for each function supported is first generated based on the operations extracted. The code includes functionality for fetching function argument(s) from the input, executing the function, and returning the result(s) to the output. When multiple functions are implemented by the actor, a scheduler is added around the core code for the functions. The scheduler checks each input for function calls. When a function call is present, it branches to the appropriate core code. After the core code is completed, it returns back to checking inputs for function calls. Currently, the scheduler is simplified. It will not execute multiple functions simultaneously although the hardware may allow it. Also it will not execute another incoming function call if the current executing function call is making a function call to another actor and waiting for the function to return. This behavior implies that the high level design in the memory domain cannot have a loop. The presence of a loop would give the possibility that the actor can directly or indirectly make a function call on itself while it is already executing a function. With our scheduler, this will result in a stall. In the future we will add the ability to execute multiple functions concurrently.

5.5.2   Linking the Translated Memory Elements.   After having translated the memory elements, they are linked together. The linking is done by implementing the semantics of the corresponding relation in the memory domain. As stated earlier, a relation in the memory domain connects an output port to an input port. The communication starts from the output port, through the relation, and goes to the input port. The token communicated is a function call request. The actor containing the receiving input port is to execute the function and returns the result through the same port and relation to the actor containing the initiating output port.

Wire Generation.   To implement a relation, for every function that can go through a relation, we generate a set of wires going from the calling actor to the receiving actor. These wires are for initiating the function call. Then we generate a set of wires going from the receiving actor to the calling actor. These wires are for returning the function call. During the wire generation for the calling side, wires for control plus one wire for each argument of the function are generated. During the wire generation for the return side, wires for control plus one wire for each returning result are generated. These generated wires are used to connect to the I/O-interfaces of the corresponding generated PEs. Like the I/O-interface wires, the generated wires are also polymorphic.

The wire generation or expansion described above is carried out for every function that can go through the relation. To determine these functions, the input and output port connected to the relation are examined. Each port has a list of functions that can go through. Here, the output port function list of the sender must be a subset of the input port function list of the receiver, since the input port must allow all the functions called by the output port.

In the memory domain, an actor making a function call, actor A, is not required to wait for the callee, actor B, to return. Actor A can potentially execute another call during that time. During this additional function execution, actor A may call B again through the same relation with the same function. Although this scenario is not possible in the current translation due to limitation in the software scheduler as discussed earlier, in general it is a valid option. This may also occur in some custom translation where actor A may issue multiple function calls before waiting for function returns. To deal with the possibility of multiple function calls coming in through a relation, the wires generated for the relations can be buffered with queues.

Buffering.   The linking process has a choice between buffering and not-buffering the relation wires. This results in a different set of control wires being generated in each case. In the case of non-buffered wires, for each calling and returning direction of a function, a valid signal wire is generated for the sending side and a done signal wire is generated for the receiving side. The valid signal is asserted to indicate that data is being sent. The done signal is asserted to indicate the data has been received. In the case of queue-buffered wires, more complex control signals are added to both the sender and receiver to interface with the queue. The queue contains two control signals on the input side, push and full. The push input is to indicate a write to the tail of the queue, and the full output is to indicate if the queue is full. On the output side, the queue contains two control signals as well, pop and empty. The pop input is to indicate a read and discard of the top element in the queue. The empty output is to indicate if the queue is empty.

The choice to buffer the relation wires affects how the control signals are generated in the linking process. Consequently, it also affects the I/O generation of the PE generation described earlier. The control signals created during the I/O generation is made to match those generated in the linking process.

5.5.3   Cache Translation Example.   Figure 4.15 shows the two-level memory hierarchy design shown in Figure 4.12 after going through the translation process automatically with the relations being buffered with queues. The cache is organized direct-mapped with a block size of four data elements and a total size of 1k data elements. It is configured to implement write-allocate with write-back capabilities, and the replacement policy is random. In Figure 4.15, the second PE from the left in the first row of PEs is the translated cache. It supports read and write functions. During a cache miss, it calls read or/and write functions on the memory. The PE furthest to the right is the translated memory. It supports the read and write functions. The rest of the PEs are the queues inserted between the cache and memory to buffer read and write functions from the cache to the memory. The third column of queue PEs buffer

further function calls from the cache to the memory. Two queues here are used
to buffer the address and data of the write function. A single queue is used to
buffer the address of the read function. The first column of queue PEs buffers
the returning results from the memory to the cache. Here we return the original
argument of the function as well. Two queues are used to buffer the address and
returning data of the read function. Two queues are used to buffer the address
and data of the write function.

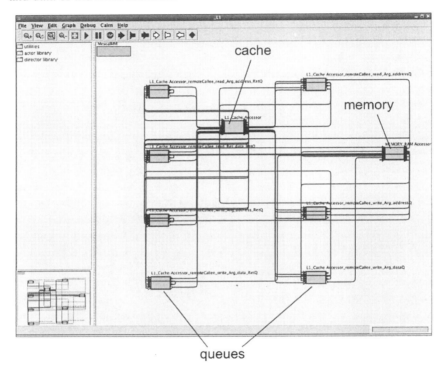

Figure 4.15.   Two-level memory hierarchy in the multi-PE domain.

   Inside the cache PE, the generated datapath contains logic to read and write
a local cache entry, detect a cache miss, allocate a new cache entry, flush a
cache entry, and fetch a cache entry from memory. The program generated
for the cache PE scans the read and write function input ports in a round-robin
fashion for incoming function calls. Upon detecting an incoming function call,
it accesses the cache for the appropriate operation and monitors for a cache
miss. If the cache access is a hit, the program returns the appropriate value
and continues with the scanning loop. If a cache miss occurs the program tries
to bring in the appropriate entry from the memory and then tries to access the
cache again. This may involve flushing out an existing cache entry and bring in

a new entry from the memory. The exact operation depends on the state of the cache. Inside the memory PE, the generated datapath contains logic to read and write an array of memory simultaneously in a single cycle. No program is need / generated here. The generation process described here is transparent from the designer. The required specifications are contained in the corresponding memory library elements.

5.5.4   Simulator Generation.

The simulator generation stage is the last element in the flow starting from the top high-level design, see the memory design flow in Figure 4.13. In this stage, a C-simulator is generated from the low-level design described in the multi-PE domain. The generated simulator preserves the semantics of the low-level design. This includes the cycle-by-cycle, bit-true execution of each PE in the design and the buffered communication between the PEs. This simulator uses individual simulators representing the different PEs in the design. It can be used to simulate a complete system, with different PEs modeling different parts the system. The generation of this simulator leverages the efficient techniques described for the Tipi simulator generation (see Section 3.2) for each processing element.

The simulator generation stage is composed of two parts. First, simulator cores are generated for each PE in the design using Tipi. Then the separate simulator cores are coupled with supporting code to form the final simulator. A similar technique can be used to generate an HDL description for the whole system. In the following sections, we describe the two parts of the simulator generation stage.

C-Simulator Generation from Multi-PE Description.

The first step in the simulator generation is to generate simulators for each PE following the Tipi flow. In this step, we start by performing type resolution across the entire multi-PE design. During the resolution, the data types for all wires with undetermined types are resolved by propagating known data types across the actors. This is similar to the intra-PE type resolution, except the type propagation continues across PE boundaries and into other connected PEs. After the type resolution, the bit widths for all wires in the design are known. Then, for each type-resolved PE in the design, we leverage the interpretive Tipi simulator generation technique for each PE. We choose the interpretive technique because it executes the processor cycle-by-cycle. As shown later, this simplifies the scheduler implementation that coordinates the PE simulator cores.

After having generated the simulator cores for each PE, the cores are combined with additional code to form the final simulator. In this process, temporary storage variables are first added for the inter-PE relations. Since the inter-PE relations are implicitly buffered, we add two temporary storage variables for each inter-PE relation. The first variable stores the relation value before the cycle, and the second variable stores the new relation value. During the cycle, any access to the relation is supplied with the old value. At the end of every cycle, the first variable is updated with the new relation value. In addition to the variables, initialization code is added to initialize the system state using the initiation techniques from the single-PE simulator generation. Lastly, a scheduler is added. The scheduler is a global entity that orchestrates the execution of the individual PE simulator cores. For every cycle, the scheduler executes each PE simulation core for one cycle. For a PE with a slower clock defined by an integer factor n, execution occurs only once every n cycles. At the end of a cycle, the inter PE relation variables are updated with the new values. In this way, scheduling of simulation events within a cycle among PE simulation cores is avoided.

As with individual Tipi simulators, probe elements can be used to monitor the inner state of the design, and stimuli files can be used to drive unconnected ports.

Extensions. Using techniques similar to the simulator generation, HDL hardware descriptions can also be generated from the multi-PE domain for the whole design. First, type resolution is carried out on all relations across the entire design. Then the HDL generation for single Tipi PEs is leveraged to generate the PE cores. Finally, the cores are connected by adding the inter-PE connections. In contrast to the C-simulator, HDL generation may be less useful for memory systems because many memory elements are often custom-built for efficiency and not synthesized from HDL. Nonetheless, the HDL hardware can still be used as a specification for the custom memory hardware.

The generation of simulator and HDL saves a large amount of time during the architecture exploration process. However, for scenarios where long simulation time is required, the designer may wish to further reduce the evaluation time by trading off accuracy versus time. This can be accomplished by evaluating the design symbolically with the use of analytical models. Similar techniques may be used to generate an analytical model of the whole system. First analytical models are generated for each PE. Then the models may be combined with additional glue logic/wrappers.

5.6   Related Memory Design Frameworks

Memory evaluation frameworks that are based on simulation can be categorized into trace-driven and execution-driven methods. Trace-driven simulators include Dinero [97] and cheetah [228]. In general, trace-driven memory simulation is less accurate than a full system simulation because the memory system is investigated in isolation. Therefore the simulation does not reflect how the memory performance affects the overall performance of the complete system. However, because only the memory system is modeled, the complexity of the simulation is smaller and easier to setup. A survey of trace-driven methods can be found in [235].

In the execution-driven method, the processor(s) driving the memory system is/are also modeled and simulated. Often the framework is designed to simulate a complete system. It may also model buses, I/O peripherals, networks, and so on. Consequently, the simulation result is more realistic. Examples of execution-driven simulators include SimOS [202], simics [156], RSIM [179], and EXPRESSION [93]. In Tipi, both approaches can be used. The designer can decide which method is more appropriate for a design scenario.

Another way to categorize memory evaluation frameworks is based on whether they evaluate a uniprocessor memory hierarchy or a multiprocessor shared memory system. Examples of uniprocessor memory simulators include cheetah, Dinero, and EXPRESSION. Examples of multiprocessor memory simulators include SimOS, RSIM, SMPCache [201], and simics. In Tipi, uniprocessor memory simulation is fully supported. Tipi is also capable of multiprocessor memory simulation, although the library does not currently contain any pre-built shared memory modules.

Most of memory simulators concentrate on cache-centric memory systems. In terms of flexibility, they allow the designer to select from a limited set of hierarchy structures and then set the parameters for each cache element in the hierarchy. In Tipi, the flexibility is greater because Tipi is designed to model a wider range of memory systems. In addition to cache configurations, the designer can construct any hierarchy structure with any connectivity and add any custom unit to the memory system. The most complete framework comparable to Tipi is based on EXPRESSION [163]. Compared to EXPRESSION, memory elements for a Tipi memory system are data-polymorphic and can thus be reused for different application-specific data types.

6.   Designing Interconnects in Tipi

In order to describe a programmable platform completely, apart from describing computation cores and memories, we also need to model interconnects, e.g. buses or network-on-chip switching elements. The major parameters of on-chip communication schemes are:

- Topology – flat or hierarchical.

- Connectivity – point-to-point or shared.

- Link type – bit width, throughput and sharing mechanisms.

- Programmability – hardwired, configurable or programmable.

Tipi's Multi-processing element (PE) view portrays the system architecture at the coarsest level of granularity. We have introduced this view in the context of memory subsystem design in Subsection 5.4. The applicability of this view, however, is more general. In this view, the basic blocks can be processing elements, peripherals, co-processors, and interconnect hardware components such as buses and switches, where all blocks are designed in the Tipi design flow described in Section 3. Architects construct network topologies using an extendible library of, for instance, network-on-chip [22, 216] components and buses.

Communication components can also be designed in Tipi. A communication component can be programmable, e.g. in order to implement a configurable switching element or a network-on-chip protocol layer. A static bus can also be described in Tipi. In this case, also the control logic for the datapath is explicitly specified so that no operations are extracted. Programmability of the communication architecture is an important design choice. Flexibility of communications is the distinguishing characteristic between systems that are highly tuned for one application domain and those that are useful for a range of application domains. For example, an architect may choose to replace hardwired bus interface units with programmable communication assist co-processors.

Like the architecture and memory views, the multi-PE view is based on a formal model of computation that gives the model functional semantics. This provides two major benefits. First, it enables a correct-by-construction synthesis path from the complete multiprocessor architecture model to an implementation, such as a synthesizable HDL model and a compiled-code cycle-accurate simulator. One way to achieve this has been explained in the context of memory subsystem design in Subsection 5.5.4. Second, the model of computation makes the capabilities of the communication architecture explicit. With this knowledge Tipi can be extended to export the correct *programming model* for the architecture to the application designer. We will briefly elaborate on this capability in Chapter 6 and like to point out that a general discussion of this topic is beyond the scope of this book.

7.   Designing Co-processors and Peripherals in Tipi

From the preceding sections it should now be clear that Tipi's architecture view can be used to describe a wide range of computation circuits, be it hardwired, configurable or programmable. Only the connection to surrounding processing elements determines whether the described ASIP is used as a main processing core, a coprocessor or as a peripheral. As reviewed in Chapter 3 in Section 2.2, peripherals can become complex, containing their own buffer memories and co-processors. So, the design of a peripheral may already employ several views from Tipi, ranging from the architecture view to specify the computation functionality, e.g. required by a protocol, the memory view to specify FIFO queues, and the multi-PE view to connect queues and processing cores together.

8. Conclusion

In this chapter, we have looked at the disciplined and efficient description and evaluation of ASIPs. We have derived three principles that design tools have to implement in order to increase design productivity and face the design challenges of modern programmable platforms:

- *Strict models* that only allow design entry of valid designs.

- *Separation of concerns*, e.g. datapath layout independent of the required control logic.

- *Correct-by-construction design* in order to avoid tedious verification steps.

We have presented the Tipi design flow that realizes these principles and improves the current state-of-the-art by obviating the need for designers to specify control and instruction sets. The designers can thus focus on the design of the datapath. From descriptions of architectures in our framework, we can automatically extract the control and instruction set, generate bit-true, cycle-accurate interpretive and compiled-code simulators, and generate synthesizable RTL Verilog®.

We have shown how our operation-level abstraction enables a semantically consistent *multi-view operation-level design* methodology. This in turn enables automatic high-speed cycle-accurate simulator, assembler, and synthesizable RTL hardware generation. Our case study with an industrial ASIC designer has convinced us of the soundness and utility of the approach. We have also shown how the Tipi design approach can be used to efficiently describe and evaluate other building blocks of a platform, such as the memory subsystem, buses, and peripherals. We believe that our methodology benefits from the fact that it supports a very broad range of programmable architectures. The key is to use a single common model to generate all needed views of the design. In this way, semantic consistency is guaranteed.

Our results have shown that our simulators are one to two orders of magnitude faster than and equivalent NC-Verilog® simulation of our generated Verilog®. Furthermore, our simulators are an order of magnitude faster than existing simulators using similar generation techniques. We therefore believe that our correct-by-construction approach to the design of ASIPs greatly increases the designer's productivity when the instruction set is not known. We also believe that the design productivity is increased by the degree of reuse and retargetability made available in our system. We are finally convinced that a path to implementation must exist, and we provide such a path with our hardware view. In this way, we help to automate the design and abstraction of a new generation of irregular ASIP architectures and whole programmable platforms.

Chapter 5

COMPREHENSIVELY EXPLORING
THE DESIGN SPACE

Matthias Gries and Yujia Jin
University of California at Berkeley
Electronics Research Laboratory

Design space exploration (DSE) is an iterative procedure to walk through the space of possible designs and find an optimal solution according to some well-defined objectives. The DSE problem is considered to be two orthogonal issues:

- How can a single design point be evaluated?

- How can the design space be covered during the exploration process?

The latter question arises since an exhaustive exploration of the design space by evaluating every possible design point is usually prohibitive due to the sheer size of the design space. The chapter gives a comprehensive overview of existing methods for DSE. This can be classified into methods for evaluating, covering, and pruning the design space. The established Y-chart scheme (Fig. 5.1) serves as a basis for illustrating DSE techniques. The Y-chart implements the separation-of-concerns principle by keeping specifications of the application (benchmark) and architecture clearly separated. A systematic mapping between the two aspects can then drive an automated exploration of the design space. Trade-offs between cost functions and optimization strategies are revealed.

We continue with a description of the Tipi architecture development system in this context. Tipi prefers a guided search of the design space for ASIPs, where the designer starts from an initial design and refines the architecture iteratively. By offering the generation of RT-level Verilog®, multi-objective search can be supported, e.g. in terms of execution time and silicon area. The memory subsystem of a programmable platform can be explored independently of the computation part. The design space is pruned by analytical models and the set of remaining designs is explored exhaustively. We also propose a method for adaptively choosing exploration algorithms according to constraints on evaluation resources, such as the maximum execution time.

In the following section we introduce some basic terminology and illustrate why DSE is a hard optimization problem. In Section 2, we compare single-objective with multi-objective optimization in general. Section 3 lists common objectives and cost functions. Section 4 reviews methods used for evaluating the quality of a single design point, whereas Section 5 surveys methods for walking through the design space. This discussion also includes methods for pruning the design space. Section 6 finally shows, how the design space for ASIPs and memory subsystems can be explored comprehensively in our Tipi design framework.

1.   Introduction

The term "design space exploration" has its origins in the context of logic synthesis. Clearly, a circuit can be made faster by spending more parallel gates for a given problem description (providing that the description offers enough parallelism) at the expense of area overhead. By extensively playing around with synthesis constraints, designers have been able to generate a delay-area trade-off curve in the design space defined by speed and area costs. This process of systematically altering design parameters has been recognized as an exploration of the design space.

1.1   Separation of Concerns for Automated Design Space Exploration

Design space exploration tasks often have to deal with high-level synthesis problems, such as the automation of resource allocation, binding of computation and communication to resources, and scheduling of operations, for varying design constraints, given a fixed problem description. In order to support early design decisions and manage increasing design complexity, exploration tasks are more often being performed on the system level. A systematic exploration often follows the Y-chart approach [127], also known as separation of concerns [126, 17], see Fig. 5.1, where one or several descriptions of the application (including workload, computation and communication tasks) and one architecture specification are kept separately. An explicit mapping step binds application tasks to architecture building blocks. The following evaluation of the mapping, e.g. in terms of performance, area, or power consumption, may require synthesis steps of the architecture description, rewriting/adapting application code, and dedicated compilation phases of the application onto the architecture in order to perform (possibly simulated) test runs. Constraints from the architecture, application, and workload descriptions may influence the evaluation. Results from the evaluation may trigger further iterations of the

mapping by adapting the description of the application and workload, the specification and allocation of the architecture (meaning the selection of architecture
building blocks), or the mapping strategy itself.

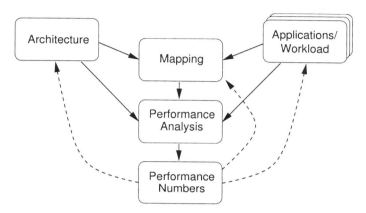

Figure 5.1. Separation of concerns/Y-chart for design space exploration.

In this chapter we will in particular focus on two kinds of methods:

- Methods that deal with the evaluation of a single design, represented by
  the performance analysis step in the Y-chart.

- Methods for the coverage of the design space by systematically modifying the mapping and the allocation of resources, corresponding to the
  feedback paths from the analysis to the mapping and architecture representations in the Y-chart.

Methods for covering the design space alter the description of the application
to adapt or refine the description according to the facilities of the allocated
architecture building blocks in order to ease a feasible mapping. During a design space exploration run, the functionality of the application usually remains
unchanged and only the workload imposed by the application may vary.

1.2   Evaluating Designs

Methods for evaluating a single design range from purely analytical methods, which can be processed symbolically, to cycle-accurate and RTL-level
simulations which need complex executable models of the design under evaluation. Before a design can be evaluated, compilation and synthesis steps may
be required, e.g. the hardware part of the design may be synthesized on an
FPGA-based prototype. The complexity of validation phases can be reduced
by correct-by-construction synthesis steps that guarantee correct implementations of the specification, thus avoiding validation by, for instance, simulation

of test stimuli. In this case, the system evaluation using stimuli can focus on extracting design characteristics only, such as resource utilization.

1.3   Complexity of the Exploration Process

The solution space for a system-level design space exploration will quickly become large if arbitrary allocations and mappings are allowed. As a simple example, assume that b distinct hardware building blocks have been allocated and communication between these blocks is not a bottleneck. The application description may consist of t computation tasks. The building blocks are of a general-purpose type, such as CPUs with different micro-architectures. As a consequence, each task can potentially be mapped onto every hardware resource, leading to b^t feasible mapping choices. Thus, an exhaustive evaluation of all possible mappings quickly becomes intractable. Consequently, there is the need for automated and disciplined approaches to reveal a representative characterization of the design space without searching the design space extensively.

The complexity increases even further if multiple objectives are subject to the search. In order to evaluate one design whether it is Pareto-optimal (see Def. 5.4 in Section 2) with respect to a set of solutions, all objective values of the design must exhaustively be compared with the corresponding objective values of every other design in the set. Fortunately, multi-objective explorations are usually bound to two or three objectives only, such as speed, cost, and power.

It should be noted that exploration methods can work either on the problem space or the solution/objective space of the design. A system based on pre-designed IP-blocks can be optimized by, e.g., searching all possible combinations of parameters exported to the designer in the problem space, such as cache sizes and the clock frequency. Those parameters are part of the initial problem specification. In contrast to that, high-level synthesis methods are driven by constraints in the solution space, such as overall latency, power dissipation, and chip area.

DEFINITION 5.1 (PROBLEM SPACE) *The problem space is defined by properties of the system that do not represent immediate design objectives but rather natural characteristics of the design space. In the context of DSE, the dimensions of the problem space often coincide with the axes of the architecture design space and may additionally include properties of the workload.*

A memory architecture, for instance, could be described by the required number of cache levels, the sizes of each level, and the caching algorithm used. An application could be represented by a task graph and an event model that triggers the tasks. All these specifications are part of the problem description and do not give any insights into primary objectives, such as the speed of the system. An exploration algorithm working on the problem space thus systematically chooses a system configuration, evaluates it, and decides whether this configuration is feasible. Finally, an optimal solution is selected in terms of one or more primary objectives, such as speed.

DEFINITION 5.2 (SOLUTION/OBJECTIVE SPACE) *The solution space is defined by the primary objectives of the design space exploration, such as system cost, speed, and power dissipation.*

An exploration algorithm working on the solution space systematically constrains feasible designs in terms of primary objectives (see Section 3), i.e. the algorithm has to determine whether a suitable design can be found that fulfills the constraints while optimizing other objectives. Design parameters in the problem space are chosen accordingly, e.g. by logic synthesis algorithms.

In the following sections, methods for evaluating and comprehensively exploring the design space are reviewed. We first discuss optimization strategies and objectives of architecture exploration.

2. Optimization Strategy

For the classification of methods we need the concept of Pareto optimality [181] which is introduced next. This property only has a meaning if a multi-objective search of the design space is performed. In the area of microarchitecture design, objectives can be the minimization of costs, power consumption, or the maximization of the speed. These objectives may show tight connections between each other. Thus, optimizing with a single objective in mind may reveal severe trade-offs with respect to the other objectives.

DEFINITION 5.3 (PARETO CRITERION FOR DOMINANCE) *Given k objectives to be minimized without loss of generality and two solutions (designs) A and B with values $(a_0, a_1, \ldots, a_{k-1})$ and $(b_0, b_1, \ldots, b_{k-1})$ for all objectives, respectively, solution A dominates solution B if and only if*

$$\forall_{0 \leq i < k} i : a_i \leq b_i \quad and \quad \exists_{0 \leq j < k} j : a_j < b_j .$$

That means, a superior solution is at least better in one objective while being at least the same in all other objectives. A more rigorous definition of *strict dominance* requires A to be better in all objectives compared to B, whereas the less strong definition of *weak dominance* only requires the condition $\forall_{0 \leq i < k} i : a_i \leq b_i$.

DEFINITION 5.4 (PARETO-OPTIMAL SOLUTION) *A solution is called* Pareto-optimal *if it is not dominated by any other solution. Non-dominated solutions form a* Pareto-optimal set *in which neither of the solutions is dominated by any other solution in the set.*

That means, designs in the Pareto-optimal set cannot be ordered using Def. 5.3. Thus, all elements in the set define reasonable solutions and they must be subject to further decision constraints in order to choose a design for a given problem. An example is visualized in Fig. 5.2. The two-dimensional design space is defined by cost and execution time of a design, both to be minimized. Six designs are marked together with the region of the design space that they dominate. Designs 1, 4, 5, and 6 are Pareto-optimal designs, whereas design 2 is dominated by design 4 and design 3 by all other designs, respectively. Without further insights into the design problem all designs in the set {1,4,5,6} represent reasonable solutions.

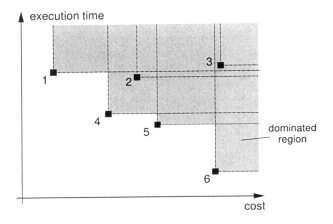

Figure 5.2.   Two-dimensional design space with Pareto-optimal designs 1, 4, 5, and 6.

Optimization methods can now be classified according to the following criteria (see [101] and the references therein):

- *Decision making before search:* The designer decides how to aggregate different objectives into a single objective (cost) function before the actual search is performed. In this way, well-established optimization methods can be applied. However, by using a weighted sum of objectives, certain regions of the design space may no longer be reachable by the search method (see the example below). Another procedure would be to convert certain objectives into constraints for an optimization problem with a reduced number of objectives. A non-arbitrary aggregation of objectives requires some knowledge about the design space to find a solution which is not sub-optimal. This might be inconsistent with one of the major goals of DSE: The determination of characteristics of the design space.

- *Search before decision making:* The search for optimal solutions is performed with multiple objectives in mind that are kept separate during the search. The result of the search is a set of Pareto-optimal solutions. Only after the search, additional criteria or preferences are applied to find an optimal solution for a given problem. In this way, an unbiased search can be done and problem-specific decisions only require the set of solutions. Hence, a single search may serve several problem-specific decisions (no rerun of the search required).

- *Decision making during search:* This category is a mixture of the two preceding groups. Here, initial search steps may be used to further constrain the design space and/or guide the search to certain regions of the design space. These steps may be repeated iteratively. Constraints and/ or guidance can be derived automatically or interactively by presenting intermediate search results to the designer.

The choice of a single- or a multi-objective search algorithm not only influences the point of time when design objectives are defined, but also affects the whole exploration process. Using a single-objective search, the result of the optimization is a single design point. That means, searches must be repeated with, for instance, varying weights or constraints on the objective function in order to explore the design space and generate a set of Pareto-optimal solutions. Depending on the shape of the objective function that aggregates several objectives, certain regions of the design space might not be reachable at all. In the example in Fig. 5.3, the region of feasible solutions in the design space defined by cost and latency is shaded. Designs one, two, and three represent possible Pareto-optimal solutions. Assume that a weighted sum of the objectives x and y is used as the objective function f:

$$f(x, y) = a \cdot x + b \cdot y \qquad a, b \in \mathbf{R}_0^+ \qquad (5.1)$$

with some weights a and b. The overall goal is to minimize f:

$$c \stackrel{!}{=} f(x, y)|_{min}$$

Equation 5.1 can be transformed to:

$$y = \frac{c}{b} - \frac{a}{b} \cdot x =: c' - a' \cdot x$$

and describes a straight line in $x - y$ space. As we can see in Fig. 5.3, the ratio of the weights a and b therefore defines the constant slope of the line, whereas the optimization goal to minimize f moves the line towards the origin. Two different optimization runs are shown where we were able to find the solutions one and three. It is clear from the picture that we will never be able to reach solution two with any combination of the weights a and b since the straight line will always hit the solutions one or three in order to minimize f (represented by *c*). In a similar fashion, one can also graphically show that the reduction of objectives by converting objectives into constraints in fact reduces the reachable space of solutions.

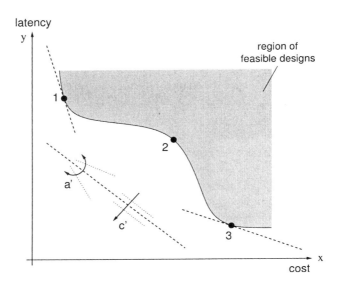

Figure 5.3.   Finding Pareto-optimal designs using a weighted sum of objectives.

Contrary to that, a multi-objective search is potentially able to find all three Pareto-optimal solutions in a single optimization run. The actual choice for one of the solutions depends on further constraints or objective functions that apply combinations of the objectives used for the search. A typical application scenario could be the exploration of an IP-core library in terms of cost and performance for a certain application domain. Suitable solutions depending on the actual application could be the fastest solution, the cheapest solution, or the solution with the best cost/performance ratio. All three solutions can be derived from the set of Pareto-optimal solutions without rerunning the search.

3.   Objective/Cost Functions and Metrics

In the following section, widely-used objectives for design space exploration are pointed out. Single-objective optimizers tend to use a weighted sum, ratio, or product of several objectives in order to consider conflicting criteria, whereas multi-objective algorithms can keep objectives separately so that the results of the search are not biased towards a certain region of the design space.

3.1   Primary Objectives

Primary objectives concern the properties of the overall system and are typically used directly as the optimization goal, i.e., they do not represent an intermediate, supportive cost metric, but drive the optimization process directly. The objectives listed in this category can universally be applied to design space exploration and are not domain-specific.

- *Cost:* The variable costs of a design could be measured as the sum of all component costs integrated in the system, e.g., based on the wholesale prices or on the manufacturing costs determined by the area consumption in the target technology and the packaging costs.

- *Power dissipation:* Optimization for power more and more becomes the focus for design space exploration. On the one hand, high-end systems optimized for speed have to cope with the generation of heat within the system that degrades the life-time of the components. On the other hand, embedded systems not only focus on the minimization of the worst-case power dissipation, but also on the power leakage during idle periods of the system in order to decrease the costs for maintenance, e.g., by extending the life-time of batteries.

- *Speed:* The speed of a design can be expressed by different metrics, such as the throughput achieved for computations and communications, the overall amount of data processed or transferred, the latency/response time for certain events (whether deadlines are met), the period length of a schedule of computations and communications, or the bare clock speed supported by the design.

- *Flexibility:* The flexibility of a design can be seen as a meta-objective since it is difficult to express this metric quantitatively. However, many fundamental design decisions are based on the need of programmability or dynamic reconfigurability in order to extend the life-time of a design, to be able to incorporate late fixes due to, for instance, changes in communication standards, or to ease the remote maintenance of an embedded system.

3.2   Secondary Objectives

Metrics in this category are either focused on the properties of only a part of the overall design or provide supportive information on the design, i.e., they reveal characteristics of the design that influence primary goals. The utilization of resources, for instance, can be seen as one component of the primary cost or power dissipation objectives. Secondary objectives are often problem-specific and facilitate the analysis of the overall design, pointing the designer to bottlenecks of the design. Common secondary objectives are:

- *Utilization of computation and communication resources:* The utilization of a resource determines the fraction of the overall execution time of a benchmark during which the resource is busy with processing the benchmark. Depending on the primary goal and the application domain, the goal of the optimization could be to maximize the utilization in order to exploit the silicon area as much as possible. Reducing the utilization however could lead to more power-efficient solutions that could also provide headroom on flexible architectures for further extensions of the application.

- *Static/dynamic profiling results:* Given an executable specification of the application, different kinds of profiling information can be extracted to guide further design decisions. An example is generating a histogram of the instructions used, e.g. data transfer vs. control vs. computation vs. bit level operations, that can indicate further exploration steps towards certain architectures supporting the most frequent operations in hardware. Profiling results can therefore be used as affinity metrics towards certain design decisions.

- *Affinity metrics:* Affinity metrics defined by Sciuto et al. [215] determine whether an executable specification of the application favors DSP, ASIC, or general purpose-like computing solutions. Exemplary metrics are the multiply-accumulate degree, I/O ratio, and bit manipulation rate of the reference application.

- *HW-SW partitioning specific metrics:* Metrics in the domain of hardware-software partitioning can be seen as a special case of affinity metrics with two design choices only. Indicators, such as potential speedups, area and communication overheads, locality and regularity of computations as defined in [189, 46], can be used for guiding the decision towards hardware or software.

- *I/O- and communication-specific metrics:* Apart from primary speed objectives, such as throughput, latency, and the number of transactions, I/O-specific metrics include the number of I/O stall cycles and arbitration penalties which affect the primary speed metrics.

- *Memory-specific metrics:* Cache characteristics that in particular represent the speed of the memory subsystem for a given application include the number of conflict and capacity misses, cache hit and miss ratios, as well as the locality of accesses extracted from the application. Metrics reflecting cost and power dissipation properties are the code size and the memory consumption of the application, e.g., generated from the maximum processing backlog of a task graph.

- *Reliability:* In the domain of embedded systems, reliability may become as important as cost and power dissipation objectives since it might be virtually impossible to service the remote system or since hard real-time functionality must be preserved under all circumstances. Reliability comes at the price of over-provisioned and/or redundant designs.

- *Deterministic behavior:* In the domain of hard real-time systems, deterministic behavior might be a primary objective in order to fulfill safety requirements. Deterministic behavior is achieved by dimensioning the design for the worst-case behavior of the system. All dynamic and sporadic events must only have bounded effects on the design, i.e., the behavior must be predictable.

- *Physical size:* The physical size and weight of a design may be of primary importance for embedded systems in the automotive domain and particularly affects the cost of the design.

- *Compatibility:* In order to partly retain the investment in previous designs, the compatibility of software, or the computing infrastructure to the new design becomes important. For the hardware part of the new design this could mean to maintain a constant interface to the surrounding computing environment, whereas compatible software requires a constant programming model for the user. Compatibility usually comes at the price of suboptimal cost and/or power dissipation objectives.

- *Usability:* Usability describes the ability of a design to ease its initialization, configuration, and programming towards the deployment in a certain application domain. Usability can also include the properties of the user interface.

- *Testability:* The support for testing a circuit can be a considerable cost factor of the design. Hardware building blocks must be included or adapted to allow a test of the circuit using externally generated test vectors. Partial or full scans must be supported by test points and scanable registers and might also require the support of certain standardized test modes, such as JTAG boundary scan. A system can also provide circuitry to generate test vectors internally, which is called built-in self test.

3.3 Combined Metrics

Single-objective optimizers combine several objectives in order to consider conflicting criteria. Multi-objective algorithms could essentially also use combined objectives in order to reduce the number of dimensions to the problem, i.e., it can make sense to only consider the speed-cost and the flexibility-cost ratios for a certain design and not speed, cost, and flexibility as separate optimization goals. The most prevalent combined objectives are:

- *Energy-delay product:* The energy-delay product is in particular used to assess embedded systems. The power requirements are traded off against the speed of the design with the overall objective to reduce the product.

- *Computations-power ratio:* This objective relates computational density to power dissipation. Designs are supposed to be better than others if they achieve more computations for a given power budget or consume less power for a given speed. The ratio between the number of computations and the power dissipation for a defined benchmark reflects this.

- *Speed-cost ratio:* This combined objective represents computational density with respect to the cost of the design. A design is better than other designs if it achieves higher speed at the same price or the same speed at a lower price. The ratio between the speed of a design and its cost combines this behavior in a single objective.

- *Flexibility-related:* In the same way the performance of a design has been combined with cost and power objectives in the preceding combined objectives, ratios and products can be defined to express the trade-off between flexibility and cost, speed, or power.

4.   Methods for Evaluating a Single Design Point

In this section, methods used to evaluate a single design point are discussed. Related work shows a variety of different approaches from detailed cycle-accurate and RTL-level simulations to purely analytical methods on relatively high abstraction levels. For some evaluation methods, the mapping step in the Y-chart [127, 17, 126] (see Fig. 5.1) to determine performance values can be fairly complex involving explicit compiling or synthesis phases, whereas other methods represent mapping decisions implicitly by varying parameter sets. The orthogonal problem of how to traverse the design space, given performance results for individual design points, will be discussed in the subsequent section.

Simulation-based evaluation can only estimate a single stimulus setting at a time, representing one particular implementation of a problem specification. The simulated workload must be chosen by the designer in a way that it represents a variety of typical working scenarios to avoid the optimization of the design for a special case. Analytical methods can help here since they are able to evaluate a design for a class of workloads (representing a range of stimuli for simulation-based tools) in a single pass. One drawback of analytical approaches, however, is that they often provide less precise results than simulation. Both evaluation techniques require a defined set of 'experimental' setups in order to produce performance-indicative, representative, reproducible, and comparable results during an exploration run (see Chapter 2 for judiciously using benchmarking).

4.1 Simulation-Based Evaluation

Simulation means to execute a model of the system under evaluation with a defined set of stimuli. A simulation can therefore only trace certain execution paths in the state space of the system that (hopefully) represent typical working modes of the design. Simulations are particularly well suited to investigate dynamic and sporadic, unforeseeable effects in the system, whereas formally verifiable systems require a deterministic behavior, given any stimuli. Results from analytical models can be too pessimistic since these models often consider the worst-case only. Simulations may reveal more realistic results for average-case optimization. One drawback of simulations is the need for an executable model. In an early phase of the design, providing such a model may impose an unsubstantiated burden for evaluating early design decisions.

4.1.1 System-Level Simulation.
System-level simulation means that the evaluation takes place at a high level of abstraction. A system is represented as an interconnection of architectural blocks of the size of whole processors, memories, and buses. The application is also represented by coarse-grained models, such as interacting processes or whole procedures.

The Ptolemy framework [145, 62] allows the designer to model and simulate the interaction of concurrent system components by using different models of computation (MoC). Through hierarchical composition and refinement the designer is able to specify software and hardware behavior at various levels of abstraction.

Lieverse et al. [154] present an architecture exploration method based on Kahn process networks. The functional behavior of an application is kept separately from models describing the timing behavior of the architecture. Applications are annotated with the computational requirement of one event. During the execution of the application model, these demands are passed on to the architecture models and evaluated. The Artemis work described in [191, 190] refines the work described in [154] in order to resolve deadlocks in Kahn process networks by introducing the concept of virtual processors and bounded buffers. A natural alteration is used in [85] to model time-dependent workloads of network processors. Here, process networks with a notion of time are employed not only to model backlog in packet queues due to limited capacity of resources and the burstiness of packet arrivals, but also to implement time-dependent schedulers, such as Weighted Fair Queuing (WFQ).

All system-level methods use some form of annotation to represent resource demands. These values may be determined by estimation, pseudo-code analysis, or even by isolated, fine-grained simulations of individual tasks to increase the accuracy at the system-level as, for instance, described in [15] and [165].

4.1.2   Cycle-Accurate Simulation.   In order to increase the accuracy of evaluating a design, an often used level of refinement is defined by the precision of a single clock cycle. Cycle-accurate evaluation means that the timing is accurately modeled on a clock cycle basis. It does not necessarily imply an accurate replication of the behavior of the system. In order to emphasize accurate modeling of timing and behavior, we find the category of cycle-accurate, bit-true simulation in literature. It means that at any given clock cycle, the state of the simulator is identical with the state of an actual implementation. Hardware models at this level of abstraction can either be based on software, modeling the timing and behavior of the hardware, or actual hardware descriptions in a hardware description language, enabling rapid prototyping on, for instance, an FPGA. The corresponding application under evaluation often is the actual application itself. The application is described in a high level programming language or assembly, and not by a functional or behavioral model of the software. A cycle-accurate evaluation thus comprises either a co-simulation of the application together with the hardware description on a hardware simulator or an integrated simulation of the application on a software model of the hardware.

The Open SystemC^{TM} [89] Initiative (OSCI[1]) tries to leverage design knowledge in the C and C++ programming languages for system level modeling and evaluation. The goal is to provide an executable specification of hardware, software, and communication parts of a design early in the design process and to establish a path of refinements towards implementation, thus bridging the gap in current design practice between high-level models and hardware description languages. SpecC [38] also supports the iterative refinement of a design and is based on the C language. SpecC is an extension of C with additional hardware and software modeling constructs, whereas SystemC^{TM} is a C++ class library. A detailed comparison of the capabilities of SystemC^{TM} and SpecC can be found in [39]. Together with the richness of the C and C++ languages however also comes the potential drawback of versatility of implementations. That means, for instance, that SystemC^{TM} models are not necessarily synthesizable. SystemC^{TM} is increasingly being used to enable heterogeneous multiprocessor simulation by encapsulating proprietary simulators with modules and using bus wrappers for coupling the modules. SystemC-based tool flows are thus particularly well-suited for exploring interconnect structures and technologies, see [21, 129, 251, 185].

Programmable processors are often investigated using cycle-accurate software models. Simulators, such as SimpleScalar [13] and SimOS [202], are spe-

---

[1]*http://www.systemc.org*

cialized in certain classes of CPUs, such as MIPS-based cores. Such a simulator therefore remodels a fixed IP-core, where only a small set of design parameters is exported to the user, such as the cache size. Modifications that affect the instruction set or variations of the datapath however require a retargetable compiler that needs to compile every application to all potential micro-architectures. These mapping and modification steps may be eased and automated using architecture description languages (ADLs). One of these ADL-based design environments is being developed in the MESCAL/Tipi project that is described in this book. Another example of ADL-based architecture exploration is the work of Mishra et al. [164, 162] based on the EXPRESSION language [93]. Further examples of ADLs are LISA [188], nML [66], MIMOLA [150], and Facile [213]. ADLs can be distinguished according to the family of architectures they are able to express (e.g. single vs. multi-threaded), the ability to integrate effects of the micro-architecture (e.g. pipelining), and their options to support automated design space exploration by, for instance, an explicit mapping step and support for retargetable code, simulation, and synthesizable hardware generation. Surveys of ADLs can be found in [195, 233].

Examples of retargetable compilers – a prerequisite for automatic, cycle-accurate design space exploration of processor micro-architectures – targeted at embedded systems [151] and ASIPs are CoWare's LISATek™ [2] [99], Chess/Checkers[3] [138], FlexWare [153], and Tensilica's XCC[4]. These compilers in particular specialize on code density and efficiency.

4.2 Combination of Simulation-based and Analytical Methods

In order to reduce the overhead involved with the simulation of a complete system under evaluation, the following methods try to reduce simulation time by gathering all characteristics, which are common between designs being evaluated and which are not subject to the design space exploration, into one initial simulation. Information extracted from the initial simulation run can be reused by all evaluations. The evaluation time narrows to the time it takes to evaluate distinctive features.

4.2.1 Trace-Based Performance Analysis.
This kind of performance estimation is in particular common for evaluating cache and memory structures, see the survey in [235]. An initial program run extracts all memory accesses and stores them in a trace. Given a cache model, the trace can then be used to calculate hit and miss statistics as well as overall performance estimates. The

---

[2] *http://www.coware.com*
[3] *http://www.retarget.com*
[4] *http://www.tensilica.com*

aim of this procedure is to save evaluation time by only doing an expensive simulation once. Different cache structures can be evaluated by reusing the same trace data collected from the initial simulation. Exemplary studies and tools that use this method for performance and energy analysis driving the design space exploration of a memory subsystem are by Fornaciari et al. [68, 69] and Givargis et al. [77].

In the work presented by Lahiri et al. [136, 134, 135] this technique is applied to the design of on-chip communication structures. An initial system-level simulation of communicating components representing the workload environment, i.e. HW and SW components surrounding the communication structure under investigation, is performed to collect traces of communications going on between components. Given these traces, different interconnect templates can be evaluated.

Zivkovic et al. [252] augment traditional traces, that usually contain information on data transfers and task executions only, with control information in order to evaluate the cost of control as well.

4.2.2 Analytical Models with Calibrating Simulation. The analytical approach described by Franklin and Wolf [71] requires an initial characterization of benchmarks using exhaustive simulation runs for a range of cache organizations. Extracted information from these runs like miss rates and load and store instruction shares are fed into analytical models which allow reasoning about resource utilization, area requirements, and performance. The approach has been extended in [72] to include power requirements.

4.3 Purely Analytical Approaches

Analytical methods come into play if deterministic or worst-case behavior is a reasonable assumption for the system under evaluation. In addition, building an executable model of the system and simulating it might be too costly or even impossible at the time of the evaluation. Analytical models thus in particular ease early design decisions by identifying corner cases of potential designs.

4.3.1 Static Profiling. Well established methods for static program analysis, such as the complexity analysis of algorithms, the dependency analysis of a static schedule of a task or function call graph to extract worst-case behavior, or a simple count of operations appearing in pseudo code, can be used for performance estimations of an application mapped onto an architecture.

Analytical approaches that take certain elements of the micro-architecture of a programmable processing core into account can be found in the domain of worst-case execution time (WCET) estimation for embedded systems. The application models usually require the absence of sporadic and non-deterministic effects. The analysis often has to neglect the impact of certain compiler optimizations. The architecture description is bounded to a single processor with simple pipeline and cache models. The formulation as an integer linear program often constitutes the core of the analysis as, for instance, described in the papers by Li et al. [152] and Theiling et al. [231].

4.3.2 Event Stream-Based Analytical Models.

For certain application domains, dedicated calculi, task, and workload models exist, which allow symbolic evaluation of a design. Richter et al. [199] (and the references therein) give an overview of mature analytical techniques for evaluating the task execution on shared resources for event streams, such as periodic events, periodic events with jitter, and sporadic preemptions. They extend these techniques by coupling the analytical models using event model interfaces, thus enabling system-level evaluations of platform-based designs. If coupling existing analytical models is not a concern, calculi provide a generalized approach to real-time embedded system design, potentially providing tighter bounds to end-to-end delays and storage bounds of shared memory implementations. One example is the network calculus [139] which has been applied to real-time embedded systems in general [44, 87]. Analytical models are well suited for early design decisions, where corner cases have to be identified quickly, maybe even automatically by using parameterized models.

4.3.3 High-Level Synthesis.

The evaluation of an application-specific architecture given a task graph may require an explicit synthesis step in order to extract area requirements and detailed timing information. Solutions to classical high-level synthesis problems of allocating resources, binding computations to resources, and scheduling operations under timing and/or resource constraints are reviewed in [183]. In the context of design space exploration, they are reapplied using exact methods, such as integer/mixed linear program formulations [29, 214], or by heuristics, such as ASAP and ALAP [30, 197], list [25, 46, 4], force-directed scheduling [59, 58], or evolutionary algorithms [25, 58] (mainly used for allocation and binding problems).

In conclusion, the trade-offs involved in choosing an appropriate evaluation method are shown in Fig. 5.4. Analytical models allow a fast evaluation of a relatively large fraction of the design space, thus enabling the identification of corner cases of the design. Over several possible steps of refinement, with increasing effort for evaluation and implementation, the design space can be bound to one particular design point. This funnel representation resembles the upper part of the platform-based design double pyramid [67], i.e., the final design point could also represent a whole platform. Methods for systematically exploring the design space on one of the layers of abstraction are discussed in the following section.

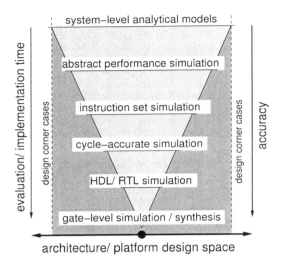

Figure 5.4. Design funnel model. Refinements of the evaluation method narrow the reachable design space (vertical direction), whereas covering algorithms explore the size of the design space (horizontal direction).

5.   Methods for Exploring the Design Space

After having discussed methods used to evaluate a single design point, this section provides a survey of algorithms used to walk through and reasonably cover the design space. Exploring the design space is an iterative process which is usually based on the Y-chart [127] approach, where concerns about functionality, architectures, and mapping are separated [17, 126]. Here, application and architecture descriptions are explicitly associated to each other in a mapping step and evaluated afterward. The mapping can include compilation and synthesis phases to enable the performance analysis. Results from the evaluation of that particular design point can then be used to further guide the exploration by varying application and architecture descriptions, as well as the mapping between the two.

In the following subsections, we provide a survey of search strategies and space pruning techniques and a list of supporting functions for automated exploration. A comprehensive survey of tools for design space exploration can be found in [86].

5.1   Strategies for Covering the Design Space

In this subsection, methods concerning the question how one could cover the design space is discussed. The mentioned categories are not strictly orthogonal to each other. In the following Subsection 5.2 methods for reducing the size of the design space are revealed that can be used in combination with any of the approaches presented in this subsection in order to decrease the exploration time.

5.1.1 Exhaustively Evaluating Every Possible Design Point. This straightforward approach evaluates every possible combination of design parameters and therefore is prohibitive for large design spaces. The design space can be reduced by limiting the range of parameters and/or by parameter quantization. Multiple objectives can easily be maintained. The search process is completely unguided and unbiased towards preferences of the designer. Examples of design systems and case studies based on exhaustive search include system-level simulation [85, 15, 250], high-level synthesis [28, 30, 47, 12, 59, 229, 214], ADL-driven approaches [162, 187], cycle-accurate simulations [128, 95], instruction set simulators [77], code parallelization and partitioning onto multi-processors [121], and last but not least trace-based analysis [136].

5.1.2 Randomly Sampling the Design Space. Evaluating only random samples is the obvious choice for coping with large design spaces. It also has the advantage of revealing an unbiased view of the characteristics of the design space. In [35], a Monte Carlo-based approach is described where random samples of the solution space are generated by randomly creating constraints for logic synthesis. Another approach is to apply simulated annealing techniques. Starting from an initial design, changes to the design become more and more unlikely with advancing search time. In [226], Srinivasan et al. compare explorations driven by simulated annealing with results using an evolutionary approach. Gajski et al. [74] combine an exhaustive search over all possible architecture allocations with a simulated annealing-based exploration of mappings for each allocation.

One could also think of combining the mentioned random walks with principles from Tabu search [81] in order to avoid evaluating the same design twice. Tabu search thus enforces diversification into unexplored regions of the design space and also incorporates mechanisms to explore around interesting design points found so far (the so-called intensification phase of the search). Tabu search however would require additional, possibly computational intensive maintenance operations in order to keep track of recent moves and bad strategic choices. Moya et al. [169] compare Tabu search with simulated annealing for an artificial two dimensional optimization scenario. Tabu search appears to be more robust against discontinuities and errors in the cost function and more effectively covers "bumpy" terrain than simulated annealing. Tabu search is also the more promising search strategy compared with simulated annealing in the architecture allocation problem investigated in [14].

5.1.3   Incorporating Knowledge of the Design Space.   Search strategies in this category try to improve the convergence behavior towards (Pareto) optimal solutions by incorporating knowledge of characteristics of the design space into the search process. The knowledge may be updated with every iteration of the search process or may be an inherent characteristic of the search algorithm itself. All mentioned methods are heuristics.

Hill climbing, for instance, evaluates the neighborhood of the current design to determine the steepest next step towards the optimization goal. In order to avoid being trapped on top of a local maximum, hill climbing requires backtracking mechanisms which might be expensive in "bumpy terrains". Moreover, the search becomes aimless on plains and is not able to recognize diagonal ridges since the probe directions would always lead to lower quality solutions. In [134], hill climbing is used to explore the mapping of communication onto channels. In [223], a kind of hill climbing is one of the investigated techniques to explore VLIW micro-architectures.

Evolutionary search algorithms combine random walk with survival-of-the-fittest ideas while constructing new generations of a set (population) of solutions. Better solutions are more likely to survive from generation to generation and new solutions can be either created by mutation (random walk) or crossover of existing solutions. Crossover tries to combine features from two good solutions to generate even better solutions. The way the designer chooses a representation of the problem and implements the mutation and crossover operations (working on those representations) also guides the search. For instance, mutation and crossover operations may generate new solutions that are not feasible. A repair mechanism could then prefer certain features of the solution over others. Thus, domain knowledge may inherently guide the search. Naïve implementations of those operations may however also avoid certain regions of the design space being reached. Moreover, problem-specific representations require recoding of the evolutionary operations for other problem domains. Single-objective evolutionary algorithms have been used in [226, 4] and multi-objective evolutionary searches are described in [232, 25, 180, 7, 8, 57, 58, 64]. Dick et al. [58] combine an evolutionary search with simulated annealing so that allocation and binding changes are less likely to happen with an increasing number of iterations.

Searches may also iteratively be guided by distance measures or other means of affinity towards certain regions of the design space. Sciuto et al. [215] define affinity metrics for applications towards mappings onto DSPs, ASICs, and general-purpose processors. Peixoto et al. [189] define metrics which favor resource sharing. Those metrics guide optimizations towards clusters of similar computations that show high locality. In this way, the communication between clusters is minimized, whereas resource sharing is maximized.

5.1.4   Path-Oriented versus Unguided Search.   This distinction emphasizes how the search progresses from iteration to iteration. Path-oriented searches are, for instance, hill climbing and evolutionary algorithms (implementing crossover). Exhaustive searches and random samples like unsupervised Monte Carlo methods belong to the class of unguided searches. Again, the latter class aims to give an unbiased view of the design space, whereas algorithms from the former class use domain knowledge of the design space to guide the search. Path-oriented walks may have the advantage of potentially reusing intermediate results of earlier design evaluations along the path. The underlying assumption here is that a design only slightly changes from one step to the next so that most of the evaluation experience from the previous design can be reused for the evaluation of the current design.

5.1.5   Single Design at a Time versus Set-Oriented Search.   This property differentiates between the number of designs that must be kept available in each iteration to perform the search. Exhaustive searches and random walks only look at one solution at a time, whereas methods exploiting domain-specific knowledge tend to use several designs at a time to find an improved design. Hill climbing and evolutionary algorithms thus belong to the latter category.

In summary, Fig. 5.5 graphically describes different search strategies for covering the design space. A discrete design space defined by two design parameters (in problem space) or two design constraints (in solution space) P_1, P_2 is assumed. Otherwise, an exhaustive search would already constitute a subsampling of the design space in this figure, since the parameters would have been quantized before the search.

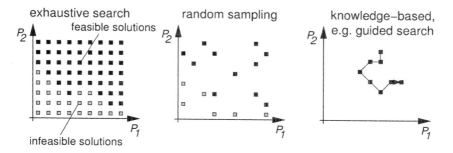

Figure 5.5.   Common approaches for covering the design space. A discrete, two-dimensional design space defined by two design parameters/ constraints P_1 and P_2 is shown.

5.2   Pruning the Design Space

All mentioned exploration methods can employ further techniques to reduce the complexity of the search by pruning the design space. Several practical approaches have been described in the literature, as will be described next.

5.2.1   Hierarchical Exploration.   Starting with a coarse problem statement, interesting regions of the design space are identified and ranked. Refined models are used to explore those regions. The higher-level models can also be back-annotated with results from the refined explorations to improve the high-level characterization of the design space. The search thus switches back and forth between high and low-level explorations.

Hekstra et al. [95] use a single simulation and profiling run of an exemplary VLIW architecture to extract timing information for all possible VLIW designs of their architecture library. Corner cases of the design are simulated to determine bounds on specific design parameters. This procedure is called *probing*. Results from probing and profiling are used to determine the design parameters which influence the solution most. Only those parameters are explored exhaustively, whereas the remaining parameters are considered with a sensitivity analysis which will be introduced later.

Mohanty et al. [165] use an analytical step first to prune the design space by symbolic constraint satisfaction. This information is used to limit the design space for trace-driven system-level simulations. Cycle and power accurate simulators may be used on the lowest level of abstraction. Results from accurate single component simulations can also be back-annotated to elements of the system simulation to improve results on a higher layer of abstraction.

In [197], different solutions of logic synthesis are explored. The problem description, e.g. a flow graph, is subdivided into templates of apparent regular structures, i.e. clusters of operations, which can be found again and again in the graph. Results from exploring those templates can then be applied to all instances of the corresponding template. A further step searches the design space at the granularity of the supergraph consisting only of templates.

Baghdadi et al. [15] use a few individual building blocks which are synthesized to RT level in order to extract timing information for possible mappings on a higher level of abstraction, i.e., this information is back-annotated to higher-level models.

5.2.2   Subsampling of the Design Space.   Subsampling the design space is a reasonable choice if the designer is interested in an unbiased exploration where an exhaustive search would be prohibitive. The subsampling pattern could be completely random, based on some regular grid, or biased by some expected shape of the design space and/or objective function(s). Monte-

Carlo based searches, simulated annealing, and evolutionary optimizers (implementing mutation) use random subsampling patterns. All approaches which quantize design parameters to reduce the design space, e.g. by allowing only a set of fixed bit widths for architecture building blocks, apply a regular sampling pattern. This property translates, for instance, to the length of one step using hill climbing. In [223], defined sweeps across the design space of VLIW architectures are used to explore the design space. Fig. 5.6 shows some common sampling patterns.

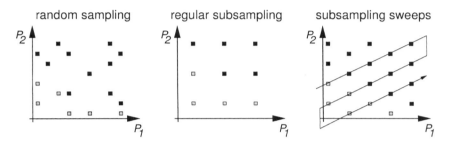

Figure 5.6. Common approaches for subsampling the design space.

5.2.3 Subdividing the Design Space into Independent Parts.
The goal of this approach is to reduce the number of possible designs by dividing the optimization problem into independent subproblems. In this way, we do not need to consider all possible combinations of design parameters but rather all combinations of Pareto solutions found for the subsystems (Fig. 5.7). Although the dimensions of the solution spaces for the different subsystems could be similar, e.g. cost and speed, the problem space of each of the subsystems usually comprises quite different, domain-specific parameters. A cache can be described by its size and organization, a VLIW core by its issue width and operator bit width, and so on.

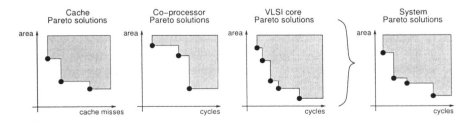

Figure 5.7. Subdividing the design space into independent parts for optimization.

Kathail et al. [123] divide the optimization of an embedded computing system into separate optimizations of the cache memory hierarchy, a customized systolic array used as a co-processor, and a VLIW processor. The exploration of memory subsystems described in [161, 132] separate optimizations concerning power consumption (by determining the cache size), main memory size (by varying the data layout), and speed (by optimizing address calculations). Givargis et al. [78] describe a method where the designer initially defines clusters of design parameters which affect and depend on each other. Separate clusters can be explored independently. Then, Pareto-optimal configurations from clusters are merged.

5.2.4   Sensitivity Analysis of Design Parameters.

The underlying assumption of this approach is the independence of design parameters. A sensitivity analysis of the design space is done using a set of reference benchmarks. In separate calibration runs only a single parameter is varied at a time, whereas all the other parameters are set to (fixed) arbitrary values. With each calibration step, the dynamic range of solution properties, such as power consumption and speed, is recorded.

Formally, given n design parameters P_i, $0 \leq i < n$ with C_i possible configurations each, the number of possible configurations C for the overall system is the product of all parameter configurations

$$C = \prod_{i=0}^{n-1} C_i.$$

In order to reduce the number of configurations for evaluation, we perform the following experiments for each parameter P_i for all r reference benchmarks b_j, $0 \leq j < r$:

- Set all other parameters P_k, $0 \leq k < n$, $k \neq i$ to arbitrary (fixed) values $P_k := P_{k_0}$.

- Evaluate the system for all possible configurations C_i of parameter P_i and note the results of interest, for instance, in terms of speed.

- The system's sensitivity S_i to parameter P_i is then defined as the difference between the maximum and the minimum result in the series of results for all configurations of parameter P_i, i.e. in this example the difference between the maximum and the minimum achievable speed by varying P_i.

Given these results for the set of reference benchmarks, the sensitivities could be averaged among all benchmarks for each parameter. The design parameters P_i can then be sorted in decreasing order of sensitivity $\{P_{S_{n-1}}, P_{S_{n-2}}, \dots, P_{S_0}\}$, where $P_{S_{n-1}}$ denotes the parameter with highest sensitivity and P_{S_0} the parameter with lowest sensitivity, respectively. The complexity of the design space exploration using these parameters can now be reduced by evaluating only designs defined by the sum of all parameter variations (versus multiplying all parameter variations in the exhaustive case) and by dropping parameters with small influence on the solution space from the exploration. For example, assuming that we stop evaluating the design space at parameter P_{S_q}, the resulting number of evaluations C' to perform for an unknown setup is given by

$$C' = \sum_{i'=S_q}^{S_{n-1}} C_{i'}.$$

Work described in [7, 8, 69, 95] uses sensitivity analysis to prune the design space. Ascia et al. [7] show one approach to extend sensitivity analysis to multiple objectives.

5.2.5 Constraining the Design Space. This straightforward task is listed as a separate point since the identification of design space constraints can form a significant initial step of a DSE run. The identification of corner cases of the design space can be done, for instance, by probing the design space or by worst-case analytical methods, such as the network calculus [139] for the network processing domain and event-stream based methods for the real-time embedded domain [199].

6. Support for Design Space Exploration in Tipi

The Tipi design flow introduced earlier in Chapter 4, Section 3 facilitates the design space exploration of micro-architectures for programmable parts, such as ASIPs. In the following, we describe how the techniques from the preceding sections can be applied for the design of ASIPs using Tipi.

6.1 Micro-Architecture Design

In Tipi, the designer lays out the datapath of the programmable processor. Control signals remain unconnected so that operations using these signals can be extracted automatically from the datapath description. The designer then constrains the set of operations and defines more complex operations by combining them spatially or temporally (assuming that no resource conflicts exist). In this way, the instruction set is defined that can be exported to the programmer at an assembly level.

6.1.1 Covering the Design Space in Tipi. The Tipi ASIP design flow favors a path-oriented and knowledge-based search of the design space. The automatic extraction of operations first checks whether operation definitions already exist. In this way, operations are found again as long as their resources in the datapath are still in place. From a designer's point of view, this means that operations will always appear under the same name as long as the corresponding part of the datapath has not been modified. This is why the Tipi approach is beneficial for a guided search. The designer starts with a general design and defines the instruction set that is composed of operations for basic functionality. At this stage, applications can already be implemented using this instruction set and the execution of these benchmarks can be profiled. The designer can then use this information in order to refine the datapath gradually with more application-specific building blocks, such as special functional units. Application-specific instructions can now be added to the instruction set and the benchmarks can be modified accordingly. Since the micro-architecture is refined iteratively, changes to the benchmarks can be kept minimal.

Tipi can of course also be used for a random and unguided search. In this case, the advantage of reusing intermediate results is lost. This means, every design point potentially has its own characteristic instruction set, and benchmarks must be implemented for each of these designs individually.

6.1.2   Evaluating a Design in Tipi.   A design point can be evaluated by using simulation. Either a simulator can be generated from the Tipi description directly, as presented in Chapter 4, Section 3.2.3, or an HDL description generated from the Tipi model can be simulated. Simulation allows the designer to determine the performance and behavior of the design. This evaluation is cycle-accurate and bit true. In addition, the HDL description can also be synthesized to a given technology in order to derive area and power consumption. The path through synthesis enables multi-objective search and constructive estimation (estimation of design quality by selectively doing real designs).

6.2   Memory Subspace

The exploration of the memory design space is split from the exploration of the ASIP core. The memory design space exploration can be automated, whereas the exploration of ASIP micro-architectures is usually guided by design experience and knowledge about the design space as a result of a disciplined analysis (see Section 3).

The approach chosen in the Tipi framework for memory exploration is hierarchical. Analytical models can be used to prune and understand the memory design space quickly. Interesting regions of the design space can then be explored by simulation. The path through simulation is again based on Tipi.

Memory is an essential part of the computer architecture. The computer performance and density is still being improved exponentially following Moore's Law [167]. In contrast to that, memory performance is being improved at a lower exponential rate than the processor core. This results in an exponentially increasing performance gap between the two architecture building blocks [248]. With this increasing performance gap, memory is becoming the bottleneck of the overall system. A thorough exploration of the memory architecture design space is an important factor in bridging this gap and eliminating the bottleneck. However, memory architecture exploration is becoming ever more difficult in the design process because of the following reasons.

■ *Complex design space:* In order to bridge the speed gap between the memory and the processing core, designers have to explore an ever larger design space to incorporate more memory architecture techniques, such as CAM and queues. This also includes more variations in memory configurations, such as more cache levels and more line size variations.

■ *Heterogeneous designs:* As the memory architecture design space becomes larger, it also becomes more heterogeneous. Designers start to explore options other than just the traditional cache technique to improve performance. For example, Intel's IXP network processor uses separately controlled SRAM and DRAM clusters and a scratchpad unit to speed up the memory aspect of network processing. The heterogeneity of the design space complicates the mapping of the application onto the hardware. Many questions arise, such as data placement and consistency. Consequently, more time and resources are needed in the exploration process to ensure a valid and efficient mapping exists.

■ *Long performance evaluation process:* After the mapping stage, it is cumbersome to evaluate the performance of each design point accurately. Often, the evaluation requires a lengthy simulation driven by traces that account for various possible usage scenarios.

■ *Shorter design cycles:* The issues listed so far all contribute to the increase in time and resources needed for the memory architecture exploration process. However, to accomplish the entire process, the designers are faced with limited resources and shorter design cycles due to fast changes in market, technology, and protocol standards.

To overcome these four issues, or obstacles, designers need to satisfy conflicting objectives. They need to explore a larger design space that is both more heterogeneous and cumbersome to evaluate accurately in less time. There are several areas of improvement that can help to ameliorate these conflicts.

■ Faster mapping methods for heterogeneous memory hardware designs.

■ Faster evaluation techniques for memory architectures without sacrificing accuracy.

- Faster memory exploration techniques with minimal sacrifice in quality of the final design. Examples are the design space decomposition used by PICO [1] and the parameter dependency technique used in [78], see also Section 5.2.

One further step is to trade off between design space exploration time and quality of the final result. Designers facing large design spaces have to cope with this trade-off already. They often use greedy heuristics, such as hill climbing and genetic algorithms, to accomplish this. However, in many cases, the trade-off in the final result cannot be well quantified.

In the Tipi memory domain, we model the time and error aspects of the design space exploration process of the memory subsystem and thus allow more explicit and quantified trade offs between the exploration time and the quality of the final result. Within our framework, the design space is specified as a multi-dimensional space. The objective function is specified along with several methods of evaluation, so-called evaluators. The evaluators' specification includes execution times and error models. A heuristic algorithm, or explorer, is selected from a group of explorers to find a good design point within the design space. The explorer selection process is viewed as a separate exploration process. This process tries to optimize the quality of the final memory exploration result by selecting the best explorer based on the design scenario. It takes into account the time and resources available for evaluation and exploration, as well as the memory exploration problem itself. In the following sections, we will discuss the memory architecture exploration formulation and the memory explorer selection in detail.

6.2.1   Memory Architecture Exploration.   In memory architecture exploration, the usual objective functions are time, power, and area. Since these objectives are complex and interdependent in nature, we follow an iterative approach to the exploration. Figure 5.8 shows a flow diagram of the exploration process. This diagram is derived from the Y-chart approach [127, 17, 126], but without the iterations on the application side. During the iterative exploration process, the application is first mapped onto a selected compatible sub-group of the design space. Afterward, the mapping is evaluated. Then the explorer analyzes the performance and determines if the optimal value is reached. If not, then the explorer selects another sub-group of the design space and restarts the loop.

Within this flow, we can now address the four obstacles discussed earlier as follows.

- We allow the designers to constrain the space, i.e. they can input their prior knowledge of the design space and prune the space accordingly.

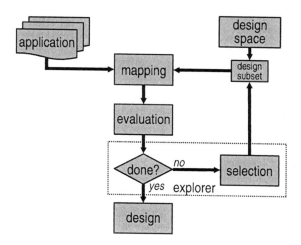

Figure 5.8. Memory exploration flow diagram.

- We keep a list of interesting evaluators that are distinguished on the basis of evaluation time and accuracy. This allows the explorer to leverage faster but less accurate evaluators when possible/needed.

- Similar to the evaluator selection, we can improve the explorer by selecting an explorer that is more likely to give good quality in the final result under the given time and resource constraints.

In the following subsections we will describe in more detail how the design space is constrained and how the time and accuracy aspects of the evaluators are modeled. Following that description, the explorer selection is discussed.

6.2.2   Constraining the Memory Design Space.

Memory Design Space.   Before discussing the issues on constraining the design space, let us first review the memory architecture design space. For an application, the memory accesses are decomposed into groups that display similar and compatible memory access patterns. This step is done manually by the designer. The design space includes all possible mappings of memory access pattern groups onto memory units. Mappings vary from all pattern groups mapped onto one memory unit to each of the pattern groups mapped onto a unique memory unit. The memory unit design space is specified by a library of high level memory configurations. Currently, this library includes stack, queue, cache, and RAM. The library can be extended to include more high-level memory configurations. Each configuration can be parameterized based on the available design choices. For example, the cache configuration includes the number of levels in the memory hierarchy, the capacity for each level, the replacement policy for each level, and so on. Further, these library elements are bit-width polymorphic and therefore can be used with any type of data width.

Parameterized Memory Design Space.   The Tipi memory design space contains parameters. These parameters may depend on each other, i.e. they do not define a multi-dimensional space on their own. For example, if all memory access pattern groups are mapped onto a single memory unit then there is no need for a second memory unit. We use the following construction to convert the memory design space into a multi-dimensional space spanned by many axes. In this construction, every design is uniquely represented by a point, or a tuple, within the multi-dimensional space. However, not all points within the space represent a valid design.

In the construction, a single axis is used to specify how memory access pattern groups are mapped onto the memory units. This mapping axis contains a finite number of points, where each point corresponds to a unique mapping. The maximum number of memory units used is the number of unique memory access pattern groups that exist within the application. For each memory unit, a set of axes, or memory unit axes, are used to specify its individual configurations. One of these axes, the so-called selection axis, is used to specify which high-level memory configuration is selected for the memory unit from the library of memory configurations. Each point on the selection axis uniquely corresponds to a high-level memory configuration, except one single point that is used to indicate the memory unit is not being used.

For each high-level memory configuration, the remaining memory unit axes are defined by the detailed parameter values for that configuration. For example, looking at caches, the set of axes include size, associativity, replacement policy, and so on. Each of these axes contains a special point, null, to indicate the scenario where the parameter is not used. In certain cases, this is an extra point in addition to all the valid values. In other cases, this point may already exist. For example, for queue sizes, the value zero corresponds to that special point, while for cache associativity, an extra point is required.

As noted previously, within the resulting parameterized multi-dimensional memory architecture design space, every design configuration is uniquely represented by a tuple. However, not all points within the space represent a valid design. Some points have invalid parameter values for the corresponding memory configuration. For example, the selection axis may indicate that the cache configuration is being used while the queue parameter values have non-zero size and the cache parameter values indicate not-in-use. To correct this, invalid points are removed from the design space by constraints. These constraints remove two types of invalid points.

- Invalid points that provide non-null values for parameters that are not in use.

- Invalid points that provide null values for parameters that are in use.

Constraining the Memory Design Space. To reduce the memory architecture design space size, which in turn will speedup the exploration process, we allow the designers to constrain the design space. Experienced designers can use their knowledge to drastically reduce the design space to a much smaller and more interesting subset. For example, for streaming applications, where caching has limited effect, the designers may eliminate the cache configuration exploration and focus on queuing techniques.

DEFINITION 5.5 (BASIC CONSTRAINT) *The simplest constraint is the basic constraint, and it is a mathematical set expression of the following type.*

$$\{design\ point\ (x_0, \ldots, x_i) \in memory\ design\ space\ | F(x_0, \ldots, x_i)\ rel\ G(x_0, \ldots, x_i)\}$$

F and G are scalar functions defined in the memory design space, and x_i's represent memory configuration parameters. F and G are built from any combination of addition, subtraction, multiplication, division, and exponentiation. The symbol rel is any relational operator belong to the set $\{=, <, >, \leq, \geq, \neq\}$. This constraint includes all design points that satisfy the Boolean expression $F(x_0, \ldots, x_i)$ rel $G(x_0, \ldots, x_i)$. Multiple constraints can be combined to form a more complex constraint. The combining is done with the set operators that include $\cap$, $\cup$, and !. The intersection, $\cap$, and union, $\cup$, operators are both binary. The inverse operator, !, is unary. As an example here is how we can eliminate the two types of invalid points in the multi-dimensional memory design space using constraints. In the following, the set expression

$$\{design\ point\ (x_0, \ldots, x_i) \in memory\ design\ space\ | F(x_0, \ldots, x_i)\ rel\ G(x_0, \ldots, x_i)\}$$

is shortened to $\{F(x_0, \ldots, x_i)\ rel\ G(x_0, \ldots, x_i)\}$ and the expression $A \Rightarrow B$ is equivalent to $(!A) \cup B$.

- To eliminate invalid points that provide non-null values for parameters that are not in use:

  When a parameter x having value i disables the usage of another parameter y, add the following constraint: $\{x = i\} \Rightarrow \{y = \text{null}\}$.

- To eliminate invalid points that provide null values for parameters that are in use:

  When a parameter x having value i enables the usage of another parameter y, add the following constraint: $\{x = i\} \Rightarrow \{y \neq \text{null}\}$.

- The above constraints are combined with the $\cap$ operators.

6.2.3   Modeling the Time and Accuracy of Evaluation Methods.   To
overcome the lengthy evaluation time required for an accurate evaluation, we
keep a list of interesting evaluators along with their models of evaluation time
and accuracy. This allows the explorer to use faster but less accurate evaluation
methods when needed. In the following, we quantify the evaluation time and
accuracy so the optimizer can make this decision.

For evaluation time, we use average single-point evaluation time. Although
this is not as accurate as it could be, we are mostly interested in a rough estimation of the total evaluation time for a set of evaluations. For this purpose and
for evaluation methods that do not have a wide distribution in evaluation time,
this is sufficient.

For accuracy, we use an error-margin random variable, δ, to model the error
range and probability for each error amount. Figure 5.9 shows the probability
distribution function, or pdf, of an example δ. Here the error ranges from +10
to -10. It is uniformly distributed, meaning any error from -10 to +10 is equally
likely.

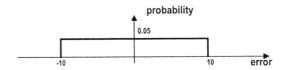

Figure 5.9.   Probability distribution function (pdf) of an error-margin random variable δ.

To acquire the average evaluation time and δ's pdf for an evaluator, we take
a reference subset of the design space and evaluate all configurations in that set.
The resulting evaluation time and accuracy can then be accounted and used to
construct the average time and δ's pdf. The larger the subset the more accurate
the result will be. This characterization step only needs to be carried out once
for similar design spaces. So its potentially large cost may be shared across
many memory exploration scenarios. In the following we give a survey of time
evaluation methods for caches as an illustrative example.

Performance Evaluation Methods for Caches.   In the case of performance
evaluation for the architecture design space of caches, evaluation techniques traditionally fall into analytical and simulation categories. In the simulation case,
an application trace is used to drive the simulation. This trace can either be collected once before the actual memory simulation or generated simultaneously
by the execution of the application program during the memory simulation.
Simulation results are usually more accurate than results from analytical evaluation techniques. However, the amount of time it takes to collect and fully

simulate the trace is usually longer than the computation time needed for analytical techniques.

There are two different techniques that are used to speedup the simulation. The first technique is to reduce the simulation detail by going to higher levels of abstraction. These simulation levels can range from cycle-accurate simulation and behavioral level simulation to transaction-level simulation.

The second technique for speeding up the simulation is to reduce the trace. The most notable of many trace reduction techniques is trace sampling. In trace sampling, only a subset, or sample, of the full trace is simulated. Statistical techniques are used to select the subset. This sampling can be done either in time or set. In time sampling [133] only the memory references within a time interval are used. In set sampling [194], also known as congruence-class sampling, only the memory references within a certain cache set are used. These two sampling techniques are not exclusive of each other. When applied, they can achieve a speedup factor of almost ten in comparison to the full trace simulation. The reported resulting error ranges from 2% - 10%.

In the analytical evaluation category, performance is modeled by mathematical equations. In most cases, the evaluation involves calculating numerical equations that return the total number of caches misses. These misses can then be translated into performance numbers once the miss penalty for the hardware configuration is given. In analytical techniques, the equations are generated from the application program. They represent a higher abstraction of the application program.

The simplest analytical evaluation is based on the stack distance curve [54]. The stack distance is defined by the number of unique memory addresses between two consecutive references to the same memory block. This distance can therefore express temporal locality in a sequence of references. For example, in the memory reference sequence $\dots A_1, B_1, C_1, B_2, A_2 \dots$ there are three unique memory addresses between the references A_1 and A_2 to memory location A, which are A, B, and C. Therefore the stack distance between A_1 and A_2 is three. There are two unique memory addresses between B_1 and B_2, which are B and C. Therefore the stack distance between B_1 and B_2 is two. The stack distance curve is a normalized distribution of stack distances within an application's memory trace. For a size x, a fully associative cache using least recently used (LRU) replacement policy, any memory reference with stack distance x or less will result in a hit. Therefore the cumulative probability under the stack distance curve from 0 to x is the hit rate for the fully associative cache of size x using LRU replacement policy. The hit rate calculated in this fashion is an upper bound for caches with less associativity. Also, since the method assumes LRU replacement policy, the result is not accurate for other replacement policies. The stack distance curve can be approximated with simple functions, such as βx^a [115], where β is a constant and α is negative [51]. This further simplifies the evaluation process. The stack distance curve method is among the simplest analytical evaluation methods. It is one of the fastest to evaluate and it has the largest error margin.

The AHH model by Agarwal et al. [2] uses a small set of parameters derived from a memory reference trace to calculate the total cache miss rate. In this model the start up and interference effects in the cache along with the non-stationary behavior of the application program are modeled. The result is more accurate than the stack distance curve method, but it takes longer to evaluate.

The cache miss equation method [76] uses linear Diophantine equations to model the memory accesses of loop-oriented scientific code. Each solution of the equations represents a cache miss. The total number of cache misses can be computed by counting all the solutions to the equations. The cache miss equation is highly accurate for scientific programs that are dominated by loops. In these cases it can achieve an almost optimal result. However, the cache miss equation does not model non-loop oriented memory access well. The computation time for cache miss equations is much larger than for the above two analytical methods. This is because counting all the possible solutions of the cache miss equations is expensive.

6.2.4   Selecting the Memory Architecture Explorer.

Within the Tipi memory exploration framework, there is a variety of exploration techniques that can be used with the design space specification and the evaluation methods provided. These include subclasses of stochastic optimization techniques, such as gradient decent, simulated annealing, and genetic algorithms. However, due to economic reasons, not all of these optimization techniques are equally beneficial for a particular design scenario. The design process itself is limited by the resources available, such as time, computation power, and engineers. While for a small, compact exploration scenario with large resources of computation and time, we can afford to exhaustively search the space and obtain the global optimal design, for large exploration scenarios with limited resources, we prefer approximations that quickly give us a good design close to the optimal design. In order to take this trade-off in accuracy and speed explicitly into consideration, the memory architecture exploration in Tipi tries to find a good explorer for the given memory architecture exploration scenario in a two-tier approach.

Two Tier Approach for Memory Architecture Exploration.   The process of finding a good explorer is framed as a separate exploration problem. Its relation with the original memory exploration problem is shown in Figure 5.10. We call this entire approach for memory exploration the *two tier approach*. In tier one, as seen in the figure on the left, is the original memory exploration. It takes the design space plus the objective function along with the evaluation methods and produces a result. Economic considerations are ignored. Tier two, seen in the figure on the right, is the explorer selection. This is a higher level exploration that takes economic constraints into consideration. Here, it takes in the original memory exploration problem in tier one, represented by the right arrow, along with the economic constraints and produces an explorer, represented by the left arrow, to be used in tier one for the memory exploration. When using the two-tier approach during the memory architecture exploration process, the memory exploration problem is first specified along with the economic constraints. Explorer selection in tier two is then carried out to select the explorer. Finally, the memory exploration in tier one is carried out with this explorer.

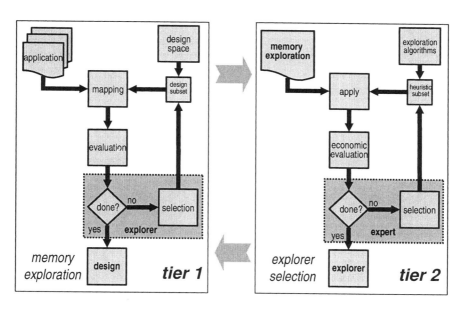

Figure 5.10. Two-tier approach for memory exploration.

The explorer selection in tier two follows a flow similar to the memory exploration in tier one. The boxes in the flow are replaced with the corresponding counterparts. The application box becomes the memory exploration. The design space becomes the exploration algorithms. The mapping stage becomes the apply stage, which tries to apply an exploration algorithm on the original memory exploration. The evaluation stage becomes the economic evaluation stage, which evaluates the economic usage and the quality of the final result produced by an exploration algorithm. The explorer becomes the expert system, which selects the explorer to be used in the memory exploration based on a set of rules.

In the following sections we will first talk about the expert system to give a high level notion on how tier two selects the explorer. Then we will talk about what exploration algorithms are considered by the expert system. Finally we will discuss how the economic evaluation is carried out on a selected set of exploration algorithms.

Expert System for the Explorer Selection.   The expert system selects the explorer algorithm that is both valid and has the highest quality in the final exploration result. It assumes that the time and quality-of-exploration numbers for each exploration algorithm are available. During the explorer selection process, every algorithm is plotted, based on its time and quality, in a two dimensional space as shown in Figure 5.11. Fast heuristics that have poor quality-of-exploration are in the lower left section. Slower algorithms with better quality are in the upper right section. Then time and quality constraints are added to this space.

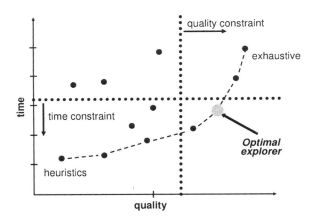

Figure 5.11.   Time versus quality trade-off for exploration algorithms.

A time constraint is limited to the form of $time \leq c$, where c is a positive constant. This is sufficient because we only need to know the maximum amount of time available for the exploration. This constraint is a horizontal line in the space that invalidates all algorithms above it.

Similarly, the quality constraint is limited to the form of $quality \geq b$, where b is a constant. This is sufficient because we only need to know the minimum acceptable quality. This is a vertical line in the space that invalidates all algorithms to the left of it. These two constraints are not necessary in all design scenarios. For scenarios where only the best achievable quality is desired, the quality constraint is not needed. For scenarios where economic resources are not an issue, the time constraint can be eliminated. After the constraint consideration, the best algorithm that is both valid and has the highest quality is selected. In the two dimensional space, this is the algorithm that is inside the valid region of space and is furthest to the right. In Figure 5.11, the best explorer is shown for some arbitrary time and quality constraints. The valid region here is the lower right section created by the two constraints.

Based on the expert system algorithm, it is clear that the best algorithm has to be on the Pareto curve of the time versus quality graph in Figure 5.11. This curve is shown in the figure as a dotted curve outlining the lower boundary of the algorithm set. Let us consider two algorithms, A and B, where algorithm A has time t_1 and quality q_1 and algorithm B has time t_2 and quality q_2, respectively. If $t_1 > t_2$ and $q_1 < q_2$, i.e., A is to the upper left of B in the time-quality space, then in all scenarios where A is a valid algorithm, B is valid as well and always better than A. Therefore A would never be chosen.

The complexity of this expert system algorithm, assuming the time and quality values are provided, is $O(n)$, where n is the number of exploration algorithms being considered in the explorer selection. Here, we assume the algorithm results are kept as a set and are not sorted in any way. The $O(n)$ complexity comes from the $O(n)$ operations required to discard the algorithms that fail the time constraints, the $O(n)$ operations required to discard the algorithms that fail the quality constraints, and the $O(n)$ operations required to find the algorithm with maximum quality within the remaining valid algorithms. Since the number of algorithms considered here is relatively small in comparison to the original tier one memory exploration problem, the time spent in the expert system is a small portion of the overall exploration time using the two-tier approach.

Exploration Algorithms. In the explorer selection in tier two, the expert system can incorporate any exploration algorithm, assuming that the corresponding time and quality values can be calculated. However, the memory exploration problem is a hard problem. Only a small subsection of all exploration algorithms may be applicable. Many exploration algorithms are only interested in finding the global optimum. There is no obvious way to reduce the quality of exploration for these algorithms. This leaves very few points for the Pareto curve examined by the expert system. In addition, the time and quality of an exploration algorithm is hard to compute in general and sometimes impossible. In the following, we discuss these difficulties in more detail and identify a class of useful exploration algorithms where we can compute these values quantitatively.

The memory exploration problem belongs to a difficult class of optimization problems. It is discrete in input space. This removes many optimization algorithms for the continuous domain. The constraints are usually non-convex. This often removes convex programming techniques, which is a large and useful class of optimization algorithms. Finally, the exact objective function is complex. Accurate evaluation requires simulation or actual execution on the hardware. It cannot be easily modeled accurately with simple mathematical equations. Consequently, very few optimization algorithms are directly applicable.

Within the small set of applicable optimization algorithms, most algorithms are designed to deterministically produce the global optimum. Examples include integer linear programming and deterministic convex programming. Only a small sub class of stochastic global optimization algorithms can trade off between time and quality of the result, and stochastic optimization problems are usually considered as one of the most difficult classes of optimization problems.

To analyze the time-quality trade off, designers often use stochastic optimization algorithms, such as genetic algorithms and gradient descent. However, for problems such as memory exploration, these algorithms only guarantee convergence to the global optimum as time approaches infinity. For real world, finite-time scenarios, these algorithms usually converge to a local optimum. There is no method to determine when the global optimum is reached or how the local optimum compares to the global optimum when only the local optimum is known.

For the explorer selection to function smoothly, we need a number of algorithms distributed evenly across the Pareto curve. The economic constraints may change in the middle of the exploration process due to various reasons, such as change in the market. Ideally, these algorithms are different stages of a single algorithm so that in the event of a change during the design process, we can dynamically tune that single algorithm by adding or reducing stages. For example, let us assume that in Figure 5.11 the points on the Pareto curve are the result of a single exploration algorithm at different stages of execution. Under the original time constraint, we need to execute to stage five to achieve the best quality. Let us assume the time constraint is reduced in the middle of the design process so that stage five is no longer valid. Normally, under this condition, we need to throw away all the intermediate results of the original explorer and start from scratch with much less time. With the multi-staged exploration algorithm, we simply reduce the execution time of the algorithm by stopping at stage four.

Hierarchical Exploration.   One class of exploration algorithms that possesses this multi-staged characteristic is the hierarchical exploration approach. With this approach, the designer would first prune the design space coarsely and find the interesting regions of space where the global optimum may reside. Then these interesting regions are pruned again at a finer detail. This pruning can continue for multiple iterations with successively finer detail until the global optimum is found. This optimization process can be stopped at any iteration of the pruning process. If the global optimum is not found during that iteration, one can still choose a point within the remaining regions and give a quality measure of it in relation to the global optimum.

Within an iteration of pruning, any algorithm can be used provided that it satisfies the following requirements. Given a set of input points and the objective function it must:

- Provide an estimation on the runtime beforehand.

- Produce an estimation on the quality for the guess beforehand.

- Produce a guess on the global optimum.

- Provide the actual quality measure of the guess.

- Provide the remaining subset after the pruning for the next iteration.

These requirements ensure that the resulting algorithms can be stacked together in a hierarchical fashion. The starting level should produce the worst quality and the more advanced levels should produce progressively better quality. For the last iteration of the hierarchical optimization, the last step of providing the remaining subset is not needed.

Several algorithms exist that can be modified to satisfy the requirements listed above. For simplicity, we focus on using exhaustive search in an iteration. Every design point is examined and any point that is unlikely to be the global optimum is discarded. We produce a class of these algorithms by coupling exhaustive search with different evaluation methods. At any iteration, exhaustive search with a faster and more accurate evaluation method is always preferred over the slower and less accurate ones. We restrict the iterations to use successively more accurate but slower evaluation methods. The earlier pruning iterations will exhaustively search with faster but less accurate evaluation methods. The following algorithm is the pseudo-code for the hierarchical exploration with variable detailed exhaustive search.

```
global_optimal_set = design_space;
evaluation_method_list;
        //initialized with n methods ordered in decreasing speed but
        //increasing accuracy
while( has_more_time()
    &&
    has_more_evaluation_methods()
    &&
    global_optimal_set_size() != 1
                                    //check if global optimum is already found
    ) {
    evaluator = next_evaluation_method();
    old_global_optimal_set = global_optimal_set;
    set_global_optimal_set_to_empty();
    for( all old_global_optimal_set) {
        evaluate();
        compare_performance();
        if(potentially_optimal()) {
            add_to_global_optimal_set();
        }
    }
}
global_optimum = guess_global_optimal(global_optimal_set);
calculate_quality(global_optimum);
```

For memory architecture exploration, the algorithms considered are all hierarchical exploration algorithms built with different combinations of memory evaluation methods, while obeying the constraint of decreasing speed and increasing accuracy of the evaluation methods for successive iterations. For n different evaluation methods that are successively slower and more accurate, there are $2^n - 1$ such exploration algorithms, which range from single iteration versions to n iteration versions. Many of these algorithms are shorter versions of the multi-iteration ones, for example, the first iteration of the two-iteration algorithm is also a one-iteration algorithm.

The hierarchical exploration described here is similar to branch and bound. In particular it closely resembles the stochastic branch and bound approach in [177]. However, the formulation of the problem is different in that we assume there are multiple evaluation methods that are at various accuracy levels. We can use these evaluators to provide the upper and lower bounds for each point in the design space. When tighter bounds are required we can simply use a more accurate evaluation method. In the stochastic branch and bound approach, measurements of the random variables are used to provide these bounds.

Economic Evaluation.   In this section, we will discuss in detail the economic evaluation for the hierarchical exploration with exhaustive search. We will first introduce probability-of-optimality and show its calculation in the case of exhaustive search. We will then define quality-of-exploration. We will show how pruning is done in each iteration along with how the global optimum is estimated. Lastly, we will show how the runtime can be estimated.

Probability of optimality for a design point is the probability of that design point being the global optimum, or being better than any other design point, in the design space. Every point in the design space has an evaluation value and a random variable modeling the error for that evaluation value. With these values, we can calculate this probability for each point. We will first derive the calculation of this probability for a point within a two point design space and then extend it to more complex cases.

DEFINITION 5.6 (PROBABILITY OF OPTIMALITY) *Without loss of generality we assume that we are minimizing the objective function in tier-one of the memory optimization. In the case of a two point design space:*

> *Let*
>> *design space $D = \{x_0, x_1\}$*
>> *the objective function, f, be a scalar function from D to $\Re$*
>> *y_0, y_1 be the evaluation value under function f for x_0, x_1, respectively*
>> *δ_0, δ_1 be the error random variable modeling the evaluation error for x_0, x_1*
>> *$Y_0 = y_0 + \delta_0$ with pdf $f_{Y_0}(y_0')$ and cdf $F_{Y_0}(y_0')$*
>> *$Y_1 = y_1 + \delta_1$ with pdf $f_{Y_1}(y_1')$ and cdf $F_{Y_1}(y_1')$*
>> *$f_{Y_0,Y_1}(y_0', y_1')$ be the joint-pdf (jpdf) of Y_0 and Y_1*
>
> *then*
>> *probability of optimality$(x_0) = P(Y_0 \leq Y_1)$*
>> *probability of optimality$(x_0) = \int_{-\infty}^{\infty} \int_{y_0'}^{\infty} f_{Y_0,Y_1}(y_0', y_1') dy_1' dy_0'$*
>
>> *we assume Y_0, Y_1 are mutually independent*
>
>> *probability of optimality$(x_0) = \int_{-\infty}^{\infty} f_{Y_0}(y_0') \int_{y_0'}^{\infty} f_{y_1}(y_1') dy_1' dy_0'$*
>> *probability of optimality$(x_0) = \int_{-\infty}^{\infty} f_{Y_0}(y_0') [1 - F_{Y_1}(y_0')] dy_0'$*

For a design space containing n points this naturally extends to the following:

$$probability\ of\ optimality(x_0) = P(Y_0 \leq Y_1, \ldots, Y_{n-1}) =$$

$$\int_{-\infty}^{\infty} f_{Y_0}(y_0') [1 - F_{Y_1}(y_0')] \ldots [1 - F_{Y_{n-1}}(y_0')] dy_0'$$

For multi-objective exploration, the probability of optimality is naturally replaced with probability of dominance. The dominance relation is defined as follows. This definition assumes that we are minimizing the multi-objective cost function.

> Given two n tuples Y_0 and Y_1 of the form $(y^0, y^1, \ldots, y^{n-1})$
> Y_0 dominates Y_1, or $Y_0 \prec Y_1$, iff $y_0^i \leq y_1^i$ for all $i \in [0, \ldots, n-1]$
> (also see Def. 5.3 and 5.4).

DEFINITION 5.7 (PROBABILITY OF DOMINANCE) *The probability of dominance between two points is defined as follows:*

> Let
> $$Y_0, Y_1 \text{ be } n \text{ tuples of the form } (y^0, y^1, \ldots, y^{n-1})$$
> then
> $$P(Y_0 \prec Y_1) = P\left[(y_0^0 \leq y_1^0) \cap (y_0^1 \leq y_1^1) \cap \ldots \cap (y_0^{n-1} \leq y_1^{n-1}) \right]$$
>
> *we assume the different objectives are mutually independent*
>
> $$P(Y_0 \prec Y_1) = P(y_0^0 \leq y_1^0) P(y_0^1 \leq y_1^1) \cdots P(y_0^{n-1} \leq y_1^{n-1})$$

Here, the probability calculation for each object component is the same as the probability of optimality for two points.

It is interesting to note that the probability of optimality forms a probability distribution over the entire design space. Therefore the sum of the probability of optimality for all points in the design space is one. Also if one point has a high value of probability of optimality, it forces all other points to have low probabilities.

Although in some cases probability of optimality may be enough to drive the iteration, for example in the scenarios where we are able to get 100% probability, in the cases where probability of optimality is less than perfect, other qualities of the design point may be more important. One example of such qualities is the maximum performance difference to the global optimum. The designers may not care about a low probability of optimality if the design point has a 5% maximum performance difference against the global optimum. In the following paragraphs, we give a general framework for calculating such qualities.

We first introduce the notion of *optimality margin*. A global optimum has an optimality margin z if and only if it is better than all other points by at least amount z.

DEFINITION 5.8 (OPTIMALITY MARGIN z) *Let Y represent the scalar objective value of a design point and x_0 be the global optimum. The optimum x_0 has an optimality margin z within an n point design space if and only if*

$$Y_0 \leq Y_1 - z, \ldots, Y_n - z$$

We can calculate the probability of having a z optimality margin with only a slight modification to the probability of optimality equations.

DEFINITION 5.9 (PROBABILITY OF OPTIMALITY MARGIN $G(z)$) *Let us denote probability of optimality margin z as $G(z)$. For design point x_0 within an n point design space, $G(z)$ is defined by*

$$G(z) = P\left(Y_0 \leq Y_1 - z, \ldots, Y_{n-1} - z\right) =$$
$$\int_{-\infty}^{\infty} f_{Y_0}\left(y_0'\right) \left[1 - F_{Y_1}\left(y_0' + z\right)\right] \cdots \left[1 - F_{Y_{n-1}}\left(y_0' + z\right)\right] dy_0'$$

$G(0)$ is the probability of optimality. $G(z)$ is a function going from one to zero. $g(z) = -\frac{dG(z)}{dz}$ is a probability distribution function, where $g(z)dz$ is the incremental probability that the point has optimality margin of exactly z. With $g(z)$ we can now define the quality of exploration as follows.

DEFINITION 5.10 (QUALITY OF EXPLORATION)

$$quality\ of\ exploration\ =\ \int_{-\infty}^{\infty} w\left(z\right) g\left(z\right) dz,$$

where $w(z)$ is a weighing function. $w(z)$ can be chosen arbitrarily.

For example, we can penalize large negative z margins and reward large positive z margins by setting $w(z)$ to be a positively sloped line crossing the origin. One can view probability of optimality as a special case of quality of exploration where the weighing function, $w(z)$, is a step function such that $w(z) = 0$ if $z < 0$ and $w(z) = 1$ if $z \geq 0$.

For the multi-objective case, we can define quality of dominance in a similar fashion. The margin Z is multi-dimensional and has the same degree as the multi-objective function.

DEFINITION 5.11 (PROBABILITY OF DOMINANCE MARGIN $H(Z)$)
We denote the probability of dominance margin Z as $H(Z)$. Given two n tuples, Y_0, Y_1 of the form $(y^0, y^1, \ldots, y^{n-1})$, $H(Z)$ is defined as follows, where Z is of the form $\left(z^0, z^1, \ldots, z^{n-1}\right)$

$$H(Z) = P\left(y_0^0 \leq y_1^0 - z^0\right) P\left(y_0^1 \leq y_1^1 - z^1\right) \cdots P\left(y_0^{n-1} \leq y_1^{n-1} - z^{n-1}\right)$$

$H(0)$ is the probability of dominance. Similarly, $H(Z)$ is a function ranging from one to zero. $h(Z) = -\frac{dH(Z)}{dZ}$ is also a probability distribution function in a multi-dimensional space. We can calculate the quality of exploration here as follows.

$$quality\ of\ exploration\ =\ \int_{-\infty}^{\infty} w\left(Z\right) g\left(Z\right) dZ$$

Here, $w(Z)$ is a multi-dimensional weighing function.

Pruning the Design Space and Selecting the Global Optimum.   During the design space pruning for a scalar objective function, a threshold of the quality of exploration is first selected. After the quality of exploration is evaluated for each design point, a design point with its quality value less than the threshold can be discarded. Alternatively, one can use a pair-wise comparison with quality of exploration calculated based on the two points being compared. A design point, x_0, is discarded if there exists another design point, x_1, such that the quality of exploration of x_0 to x_1 is less than the pre-established threshold. When a guess of the global optimum is required in the case of multiple design points still remaining, the design point with the highest quality is selected.

For multi-objective functions, design space pruning is done with pair-wise comparison with quality of exploration. A design point, x_0, is discarded if there exists another design point, x_1 such that the quality of x_1 to x_0 is more than the pre-established threshold. This is slightly different from the scalar objective case because the quality of exploration here resembles the dominance relation. Only high quality suggests one design point is better than the other. Two design points may both have low quality against the other one. Therefore, we are only interested in removing the points that are being dominated. After the pruning, the remaining design points form a fuzzy Pareto curve. As the pruning is refined to more accuracy after more iterations, the fuzziness of the Pareto curve decreases.

Runtime Calculation.   In this section, we show how the estimation of the runtime can be calculated for the hierarchical exploration with variable detailed exhaustive search. We assume the objective function is scalar. A similar technique is in development for multi-objective functions. For the calculation, we will first introduce the concept of resolution for an iteration within the hierarchical exploration. Then we will define the distribution for the differences from the global minimum of the design space. Finally, we will estimate the amount of design space pruned in each iteration and the running time for the hierarchical exploration.

The resolution, θ, of an iteration within the hierarchical exploration is the minimum difference that the iteration is able to distinguish. Therefore at the end of the iteration, all remaining design points will be within θ of the actual global optimum. To calculate the resolution for an iteration using exhaustive search we first examine the algorithm's ability to separate two points by calculating the expected quality of exploration.

DEFINITION 5.12 (EXPECTED QUALITY OF EXPLORATION) *Let x_0, x_1 be two design points and $A(x)$ be the actual objective function. $A(x_0) - A(x_1) = z$, but the actual objective function values for x_0, x_1 are not known. Let $a(x)$ be an evaluation method with error random variable δ with pdf $f(z)$*

and cdf $F(z)$. *The* expected quality of exploration *is then defined as follows.*

expected quality of exploration $(x_0) =$

$$\int_{-\infty}^{\infty} \int_{-\infty}^{\infty} f(y_0 - z) f(y_1) \int_{-\infty}^{\infty} w(z') \left[\frac{\partial \int_{-\infty}^{\infty} f(y - y_0)[1 - F(y - y_1 + z')]dy}{\partial z'} \right] dz' dy_0 dy_1$$

Here $f(y_0 - z) f(y_1)$ represents the probability of evaluation method, $a(x)$, evaluating x_0, x_1 to be y_0, y_1, respectively. We assume the two evaluations are independent of each other.

$$\int_{-\infty}^{\infty} w(z') \left[\frac{\partial \int_{-\infty}^{\infty} f(y - y_0) [1 - F(y - y_1 + z')] dy}{\partial z'} \right] dz'$$

represents the quality of exploration calculation. Intuitively, the expected quality of exploration for x_0 is zero if the non-zero region of pdf $f(z)$ has a width of ε and $z > 2\varepsilon > 0$.

DEFINITION 5.13 (RESOLUTION θ) *Given two design points with separation* z, *if the expected quality of exploration for the less favored point over the other is less than a given threshold, then on average we are be able to distinguish the two and discard the less favored point. Therefore it is natural to define the resolution as the smallest occurring* z. *The expected quality of exploration is calculated on the less favored point.*

Resolution $\theta = sup (z$ *s.t. expected quality of exploration* $(z) <$ *threshold*)·

In order to estimate the amount of pruning we need the notion of a Δ min distribution, $m(z)$.

DEFINITION 5.14 ((CUMULATIVE) Δ MIN DISTRIBUTION $m(z)$) *The* Δ *Min distribution* $m(z)$ *is the distribution of the differences from the global minimum of the entire design space.* $M(z)$ *is the cumulative version* $M(z) = \int_{-\infty}^{z} m(z) dz.$

$M(x)$ is used in estimating the amount of pruning. The actual objective values of the design space are not needed here. Therefore we do not need to evaluate the entire design space. $M(z)$ is estimated by the designers. This is an interface for the experienced designer to help the tier-two explorer selection to select the right exploration algorithm. As we will see in the following, designers only need to estimate $M(z)$ accurately at the points where z equals to the resolution values of the different iteration algorithms.

Pruning and Runtime Estimation. Given a hierarchical exploration algorithm with n iterations, from 0 to $n - 1$, where each iteration has the resolution

θ_i, and $\theta_i > \theta_j$, if $i < j$, we can estimate the pruning on a design space with the Δ min cumulative distribution, $M(x)$, in iteration i. At the end of an iteration, all remaining design points are within θ_i of the global minimum. Therefore the number of remaining design points after iteration i is $M(\theta_i)$. The amount of design points pruned in iteration i is the difference between iteration $i - 1$ and i, or $M(\theta_{i-1}) - M(\theta_i)$. In the first iteration the entire design space is present at the beginning. Therefore $\theta_{i-1} = \infty$ when $i = 0$.

When using the exhaustive algorithm in each iteration, every remaining design point at the beginning of the iteration is evaluated. Further, for every point, the quality of optimality is calculated and compared to the threshold. Therefore the runtime for iteration i is $(et_i + ct_i) M(\theta_{i-1})$, where et_i is the evaluation time for the evaluation method used in iteration i, and ct_i is the time for the quality of exploration computation and comparison.

The runtime for the entire hierarchical exploration is the sum of every individual iteration, or $\sum_{i=0}^{i=n-1}(et_i + ct_i)M(\theta_{i-1})$.

7.   Conclusion

We believe that the separation of concerns – application, architecture, and mapping specifications – is the basic principle that enables comprehensively exploring the design space. In this chapter we have shown how the different elements of our design methodology described so far play together following this principle in order to perform design space exploration (DSE). The first element of our methodology, judiciously using benchmarking in Chapter 2, deals with the application and environment specification. The second element in Chapter 3, inclusively identifying the design space, constrains the space of possible architecture specifications. The DSE problem can now be introduced as two orthogonal issues: How to evaluate a single design point, and how to walk through the design space. The former point is the main focus of the third element of our methodology described in the preceding Chapter 4, where the state-of-the-art of efficient ASIP modeling and evaluation is shown. The latter point is addressed in this chapter where we classify methods into space pruning and space covering algorithms. That means, the fourth element of our design methodology, comprehensively exploring the design space, closes the loop of specifying and evaluating designs in order to perform systematic DSE.

Using our design framework, Tipi, we have shown how we envision disciplined design space exploration should be performed. On the one hand, Tipi reduces the burden for design entry of application-specific programmable parts by automating the generation of datapath control logic. This technique is particularly beneficial if a guided, e.g. knowledge-based, search of the design space is pursued since immediate results from preceding design points can partly be reused. On the other hand, for well defined and parameterizable sub-spaces like the memory sub-system, we have illustrated how the design space can be evaluated and covered automatically. In this context, we have introduced a new method that adaptively trades off the precision of evaluation versus the time of evaluation and exploration.

The methods introduced in this chapter enable the designer to explore the design space comprehensively in a disciplined way. As the fourth element of our design methodology, we believe this is mandatory in order to face the design challenges of modern programmable platforms and avoid ad-hoc integration of previous design decisions.

Chapter 6

SUCCESSFULLY DEPLOYING THE ASIP

Niraj Shah[1], William Plishker[1], Kaushik Ravindran[1], Matthias Gries[1], Scott Weber[1], Andrew Mihal[1], Chidamber Kulkarni[1], Matthew Moskewicz[1], Christian Sauer[2], Kurt Keutzer[1]

[1] *University of California at Berkeley*
Electronics Research Laboratory

[2] *Infineon Technologies*
Corporate Research
Munich, Germany

Apart from the design flow for building the architecture of a programmable platform, the software development design flow impacts the overall time-to-market considerably. A programmable platform becomes useless if it cannot be programmed effectively. That means, on the one hand, software development tools must offer to the programmer a reasonable abstraction of the underlying hardware. On the other hand, this abstraction must still be capable of exporting critical features of the architecture so that reasonable performance can be achieved. Finally, manually reimplementing these tools for every single design alternative during design space exploration is not feasible. A tool flow has to support the automatic generation of software development tools. This chapter describes how programming models are defined in the Mescal project to enable the successful deployment of ASIPs. In addition, the automatic generation of fast cycle-accurate, bit-true simulators eases the deployment of the programmable platform since software can be developed before actual silicon is available.

We would like to note that a comprehensive discussion of all issues involved with the deployment of a programmable platform is beyond the scope of this book. In this chapter, we briefly discuss all currently available functionality in Tipi that eases the deployment of ASIPs. We recognize areas of current and future research in this field. This includes the question of how to compile from a more general application description, rather than a sequential language, down to a programmable platform. For one particular application domain – network processing – we introduce a specific programming abstraction. This abstraction, a so-called programming model, is natural for the application domain and allows the programmer to exploit the resources of the underlying network processor considerably faster than by using completely hand-written code. We also show how the cumbersome task of finding a feasible mapping of threads of execution onto processing resources can be automated. We are currently generalizing these experiences to less restricted application and architecture specifications so that compilation from heterogeneous models of computation to Tipi multiprocessor systems will be possible.

1.   Deployment Requirements on Design Tool Flow

Architecture description language (ADL) based frameworks currently comprise the best practice for generating software development tools for ASIPs. These frameworks generate several tools from a central description of the architecture and the corresponding instruction set. These include:

- *Assembler and compiler:* Apart from their role in design space exploration, these tools are required for the acceptance of a platform by the majority of application developers. Most of the algorithm development for signal processing and high performance computing has been done in sequential programming languages for decades. This situation implies that compilation from popular languages like C and C++ is a basic requirement.

- *Simulator:* Apart from its role in assessing the performance and behavior of an architecture, a fast bit-true and cycle-accurate simulator of the platform allows an application programmer to develop software before the platform is actually manufactured, i.e. without access to real hardware. If the simulator can be bundled with a compiler, software can be developed before the hardware architecture is finalized.

Tipi supplies an assembler and simulator by generating them automatically from the central description of the designed microarchitecture. In the following, we point out the specifics that distinguish Tipi from existing ADL-based solutions. In an ADL-based tool flow, the designer has to manually specify the instruction set architecture and instruction encodings, or she/he has to design the datapath control explicitly. In Tipi, we believe a designer mainly wants to focus on the design of an application-specific datapath, and that the definition of the instruction set and datapath control should be generated automatically. In the following subsections, we show how Tipi's approach simplifies the definition of an instruction set that can then be communicated to the application designer.

1.1 Support for Successfully Deploying Architectures in Tipi

In Section 3 of Chapter 4, we have discussed how the Tipi design flow is founded on the concept of extracting multiple consistent views from a single central description. In this context, we have mentioned the operation and constraints views. In this subsection, we explain how these views facilitate the deployment of ASIPs by simplifying the definition of instruction sets seen by the programmer.

1.1.1 Operation and Constraints Views.

As an illustrative example we use a simple datapath design containing two register files, an ALU, a multiplier, an adder, global memory, and a couple of pipeline registers. This datapath is visualized in Figure 6.1. The figure also shows how this design would look like in Tipi's architecture view.

In accordance with Tipi's correct-by-construction operation-level design principle, all control inputs to datapath elements are left unconnected. Tipi uses this information in order to extract all possible primitive operations of the datapath (see Chapter 4, Section 3). In Figure 6.2, we see the resulting operation view in Tipi. On the left, all primitive operations are listed. One operation is selected that initializes the first operand of an ALU operation from a register file. The operation uses one parameter that represents the read address from the register file. A text box also displays the semantics of this operation. The operation view can be used to restrict the set of primitive operations. Usually, not all primitive operations are required and these operations can safely be removed from the set of operations that are exported to the programmer.

The information displayed in the operation view is still quite fine-grained. Although these primitive operations can already be used to program the datapath, the programmer would have to think about what operations could be used concurrently and sequentially without generating conflicts in the underlying datapath. Tipi therefore provides a constraints view, where constraints on the usage (i.e., sequential and temporal combination) of primitive operations can be imposed on the set of available operations. From a designer's point of view, she/he can define the instruction set to be exported to the programmer in this view.

An example application of a constraint is displayed in Figure 6.3. A subsection of the design in Figure 6.1 is shown. In this example, two primitive operations can be combined to form a single operation since there are no spatial constraints in the datapath. This constraint states that the two primitive opera-

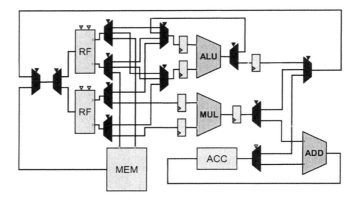

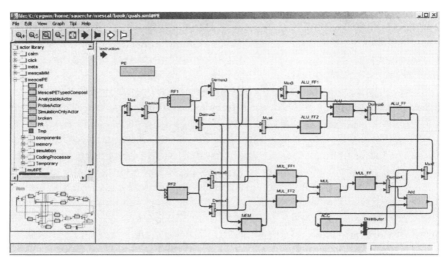

Figure 6.1.   Architecture view with simple datapath.

tions may be executed concurrently. In this case, two flip-flops are initialized from the register file to be used in a following ALU operation.

Another form of a constraint is a temporal constraint. A temporal constraint restricts the sequential usage of operations. For instance, it might make sense to only allow the initialization of the flip-flops shown in the preceding figure if the initialization is followed by an ALU operation. In Figure 6.4, an example of a temporal constraint is shown to combine the operand fetch, ALU operation, and writeback phases in one predefined sequence. For the first operation in

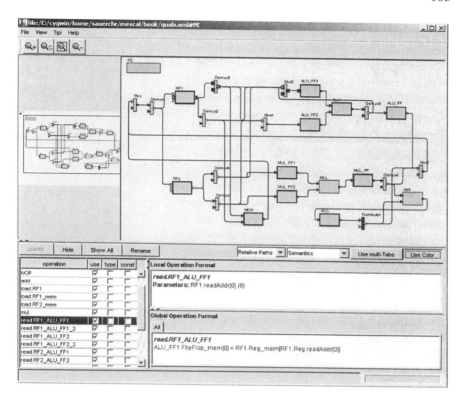

Figure 6.2. Operation view in Tipi.

this sequence, we employ the spatial constraint defined earlier in Figure 6.3. Constraints can make use of other constraints.

Figure 6.5 displays a screenshot of the constraints view in Tipi. The definition of the constraint from Figure 6.4 is shown. The constraint requires three parameters, the destination register and the two operand registers. On the left, one sees all allowed primitive operations and defined constraints. From a designer's point of view, this is the instruction set for the underlying datapath that is exported to the application developer. This instruction set comes with a corresponding assembler so that the designer never has to cope with any binary instruction encoding or datapath control logic.

We would like to add that there is a Tipi tutorial available that describes this procedure in more detail, see [88].

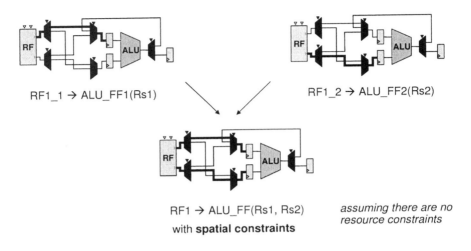

Figure 6.3.   Defining a spatial constraint to exploit concurrency of operations.

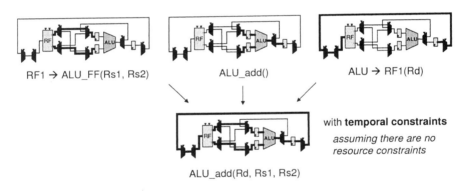

Figure 6.4.   Defining a temporal constraint to execute operations sequentially.

1.1.2   Simulator Generation.   Simulators can be generated based on this instruction set and the datapath specification. The underlying techniques have been described in Chapter 4, Section 3.2. Since Tipi's simulators are cycle-accurate and bit-true [240], they accurately allow the programmer to develop software before the hardware is finalized. For application-specific programmable parts with possibly irregular datapath structure this accuracy is mandatory. A simpler instruction set simulator would not be enough to really exploit the hardware. Our case study in Chapter 4, Section 4 has shown that Tipi-generated simulators achieve a performance that is comparable with existing

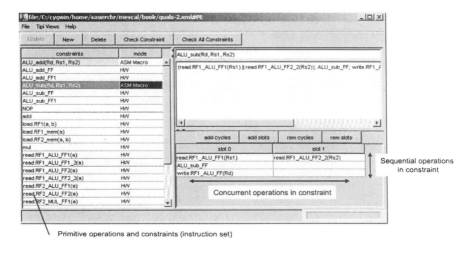

Primitive operations and constraints (instruction set)

Figure 6.5. Constraints view in Tipi.

instruction set simulators. We thus believe that Tipi simulators can indeed be used to develop real software projects concurrently to the hardware design. The constraints view can play a key role in this situation. A hardware designer may define a complex constraint consisting of several primitive operations of the underlying datapath. At a later stage, this composite operation may be replaced by a more specialized hardware implementation in the datapath. The instruction set given to the programmer can remain the same during this process.

Tipi's correct-by-construction approach of generating operations automatically from datapath descriptions particularly eases the successful deployment of ASIPs. The designer can always be sure that the defined instruction set is actually implemented by the hardware. Also, the corresponding simulator and HDL descriptions are consistent with the instruction set and datapath definition. Traditional design flows require the designer to perform additional verification steps to check for this consistency.

1.2 Improving the Programmer's View

The programming abstraction introduced so far is an instruction set architecture. ADL-based tool flows often offer compilers for sequential languages like C and C++. For application-specific processors described in Tipi we want to offer more general programming abstractions and automated compilation steps to ease the deployment of heterogeneous multiprocessor systems. For this purpose we need the notion of a *programming model* introduced next.

DEFINITION 6.1 (PROGRAMMING MODEL) *A programming model is an abstraction that exposes the subset of the details of an architecture that are necessary for a programmer to efficiently implement an application. It is a programmer's view of the architecture that balances opacity and visibility:*

> *1 Opacity: Abstract the underlying architecture*
> This obviates the need for the programmer to learn intricate details of architecture just to begin programming the device.

> *2 Visibility: Enable design space exploration of implementations*
> This allows the programmer to improve the efficiency of their implementation by trading off different design parameters. Our goal is that the full computational power of a target device should always be realizable through the programming model.

In summary, a programming model supplies an approach to harvesting the power of the platform. A programming model will inevitably balance between a programmer's two competing needs: Desire for ease of programming and the requirement for efficient implementation.

Furthermore, we recognize a broader trend of the search for application-specific solutions with fast time-to-market. This trend is drawing system designers away from the time-consuming and risky process of designing application-specific integrated circuits (ASICs) and toward programming application-specific instruction processors (ASIPs). As system designers increasingly adopt programmable platforms, we believe the programming model will be a key aspect to harnessing the power of these new architectures and allowing system designers to make the transition away from ASICs. The programming model is therefore a necessary, but not sufficient, condition of closing the implementation gap between a very general application description and an application-specific, heterogeneous platform.

We would like to note that a comprehensive discussion of how compilation is being pursued in the MESCAL project is beyond the scope of this book. In the following, we give an overview of the approaches applied for simplifying the deployment of a platform described in Tipi.

1.2.1   Compilation.   The goal of a good programming model is to provide much of the structure required to easily and efficiently program a given architecture. However, the interface that programming model exposes to the programmer must eventually be mapped onto the architecture. This mapping process is a generalization of what is often loosely referred to as compilation.

The mapping process is generally divided into multiple conceptual layers in order to deal with the complexity of mapping a heterogeneous parallel application onto a heterogeneous parallel architecture. Generally, the upper layers of the mapping process consist of high level abstractions defined using syntactic transformations. Higher level computational, communication, and control primitives are defined, and rules governing the composition of these primitives are specified. In some cases these primitives can be defined to match very closely with the architecture. Such primitives are naturally easy to extract from an architecture, and programs using them can easily be mapped. Other times, simple syntax-transformation approaches can be used to decouple the architecture from the programming primitives without too much loss of performance. This is often referred to as the library approach, and again in this case mapping generally is easily accomplished.

However, when dealing with the lower levels of the mapping process, such as scheduling and resource allocation on a cycle by cycle basis within processor datapaths, neither of these approaches is viable. Low level computational and scheduling primitives require the programmer to write assembly or microcode. This is generally not easy, scalable, or fun. Higher level primitives typically have no expressible syntactic transforms into pipeline level operations. Thus, automated mapping tools are a requirement to efficiently utilize architectural features while obeying architectural constraints.

Automated mapping tools are useful at all levels of the programming model. Typical approaches may utilize many mapping tools operating at many levels of abstraction. Thus, the cost of the creation of such tools themselves must be considered when designing programming models. Higher level mapping tools tend to be more independent of a particular architecture, or even a particular programming model. However, lower level tools tend to be more dependent on a given architecture, and thus the development of such tools can be a serious bottleneck. This is especially true if one wishes to perform design space exploration over many architectures for a given application domain and (high level) programming model. The availability of either generic or automatically generated tools for low level scheduling and allocation of architectural resources is extremely desirable.

1.2.2 Low Level Mapping Strategy.

Problem Description. Large amounts of heterogeneous low level parallelism may be present in Tipi processing elements. Typical PEs include multiple register files, pipeline registers, memory access units, stream gate units, functional units, and an arbitrary network of paths connecting them. Traditional register-to-register style ISAs cannot capture the capabilities or constraints of such an architecture, and the cycle-by-cycle microcode 'ISA' that is provided by the Tipi tools is difficult to use directly. Spatial and temporal grouping of operations provides a capability for abstraction, but it cannot easily handle the highly variable cycle-level behavior of irregularly pipelined and forwarded datapaths.

There are approaches to directly produce microcode-level instructions from computational DAGs at roughly the level of basic blocks or loop kernels. Performance critical sections can be directly scheduled and allocated onto the datapath using such methods. Then, syntactic methods can be used to glue such sections together to form program units. The resulting program units can represent a sufficiently high level of abstraction to be targeted by higher layers of the mapping. It may also be required to efficiently combine, schedule, or otherwise manipulate partially mapped program units at the behest of the higher mapping layers. Furthermore, the approach used must be easily (preferably automatically) retargetable across various Tipi architectures.

Low Level Code Generation. The current research-in-progress low level code generation method for Tipi PEs is an extension of various DAG matching, ILP-based, and brute force 'super-optimization' code generation approaches [5, 137, 166]. For a given Tipi architecture, the code generator directly reads the architecture and needs minimal or no configuration to be able to begin generating code. For the generation of straight-line microcode, generally no configuration is needed. For code containing control flow, the code generator must either be able to recognize and understand the datapath components that perform control flow on the microcode stream, or those elements must be tagged and described explicitly in the architecture.

For code generation, the following steps are performed:

- *DAG generation:* A relatively small (between 30-100 intermediate values) computational block is first isolated and converted into a DAG.

- *Constraining the DAG:* External constraints on the sources and sinks of the DAG are specified. When scheduling blocks such as loop bodies, the external constraints generally consist of reading and writing via either memory or stream type units rather than register files or pipeline registers. This is because the sources and sinks will be used for each loop iteration.

- *Scheduling the DAG:* Nondeterministic symbolic simulation of the program DAG matched against the architecture is then used to define a space of possible cycle-unrolled microcode schedules to consider.

This simulation will be circular in time for loop body blocks where control is expected to flow circularly in the common case. Determining the minimal number of cycles of simulation that are required to include some schedule that can compute a given block generally requires search. However, reasonably accurate heuristics can often guess the correct number of cycles required based on analysis of the available functional units in the datapath versus the number of matching operations in the program DAG. Since the architecture is directly unrolled, the unrolled simulation includes all of the basic structural constraints of the architecture.

To simplify the encoding of the search space, specific register allocation is deferred until after scheduling. Cardinality constraints on the contents of the register files are encoded to ensure that the resulting schedule is always allocatable. Multiple blocks can be stitched together either by generating glue code, or by manipulating the external constraints of blocks to be directly compatible. Loop bodies will generally need special preamble and postamble glue code that can only be generated after the loop body itself is code generated. In the current implementation, a customized SAT-solver is used to perform the search over the set of unrolled possible microcode traces.

1.2.3   High Level Mapping Strategy. A key feature of ASIP platforms is the presence of numerous heterogeneous programmable elements in the architecture. This architectural diversity is witnessed in the IXP1200 and numerous other programmable platforms. Deploying multiprocessor systems brings challenges above and beyond those encountered in programming individual PEs. In the MESCAL project, we are interested in building and programming systems composed of multiple heterogeneous Tipi processing elements. While a complete treatment of this subject is beyond the scope of this book, we outline the major concepts in this section to address application concurrency and mapping to heterogeneous platforms.

Facing Concurrency. First, we challenge the dominant philosophy that designers can view multiprocessor machines as a collection of separate processing elements and program each element individually. This *system-in-the-large* way of thinking is far too weak to handle the most pressing issue in embedded system design – concurrency. Without formal methods for reasoning about concurrency, designers must use complex ad hoc methods to implement application concurrency on the multiprocessor architecture.

Current best practice in programming multiprocessors is the use of APIs for message passing, POSIX® threads, or shared memory multiprocessors (e.g., OpenMP™ [1]). In all cases, the programmer is forced to explicitly manage concurrency, communication, and synchronization in a sequential programming language, such as C++ or FORTRAN. The automation of this process is only well understood for regular hardware and software structures, such as systolic arrays performing scientific programs.

---

[1] http://www.openmp.org

This complexity is compounded by the fact that modern embedded applications use many flavors of concurrency in combination – heterogeneous concurrency. With manual techniques, designers end up spending a majority of their time simply trying to achieve a functional system. There is seldom enough time to perform effective design space exploration.

To address this issue, we promote the use of *models of computation* (MoC) to capture application concurrency. A model of computation is a mathematical formalism that describes how components in a system interact. For embedded software, programmers can use combinations of models of computation to describe the different facets of the application [143] in a way that is most natural for the application domain. One example would be to describe control flow aspects with finite state machines and data flow with synchronous data flow.

In order to use combinations of MoCs, we look towards systems like Ptolemy II [145, 62]. In the MESCAL project, we have developed a variation on Ptolemy II called Cairn. Cairn preserves the core features of Ptolemy, such as models of computation and actor-oriented design. We replace Ptolemy's Java® software architecture with an actor description language based on the Tipi actor semantics language (see [239] and the examples in Chapter 4, Section 3). This gives us the ability to capture concurrent computation in a language that is formally comparable to the capabilities of Tipi architectures.

Mapping to Heterogeneous Resources.   Capturing application concurrency is a necessary prerequisite to thinking about how to implement that concurrency on the resources of a heterogeneous multiprocessor. In most applications and architectures, there is not an exact match between these types of concurrency. This problem is the *concurrency implementation gap*. For example, the number of PEs in the architecture is probably not exactly the same as the number of processes in the application. The communication architecture of the platform is probably not an exact match for the communication semantics and data transfer patterns of the application. Programmers must find a mapping between application concurrency and architecture concurrency to solve this implementation gap.

We attack this issue by providing a disciplined mapping methodology that treats concurrency as a primary concern. Programmers begin with the high-level application models that were described using models of computation. The model of computation formalisms help divide the application into control, communication, and computation facets. Each of these facets will be compared with a similar facet of the target architecture.

- The *control facet* describes how the application is composed of parallel threads and processes, and how these processes execute relative to each other. These processes will be mapped to processing elements in the architecture.

- The *communication facet* describes the semantics of how application processes communicate. This functionality will be mapped to the communication architecture of the platform.

- The *computation facet* describes the mathematics that the application performs. This will be mapped to the instruction sets of the individual PEs.

We provide a toolset that permits designers to rapidly make these mappings and understand the results. Since we use the language of Tipi operations to describe both application requirements and architectural capabilities, we can perform correct-by-construction code generation of a functional implementation. This is combined with the fast, accurate Tipi PE simulators described in Chapter 4, Section 3 to give designers rapid feedback on the quality of a mapping.

As a result, designers are not required to translate their high-level application models into a low-level language for implementation, nor are they required to create a new code generation tool chain for each flavor of processing element. These features are crucial to a methodology in order to support efficient design space exploration.

1.2.4   Conclusion.   In summary, efficiently programming a heterogeneous multiprocessor platform is more than simply programming multiple individual elements. Developers must begin by carefully modeling the application's inherent concurrency. Models of computation provide intuitive abstractions for this task. Next, developers must explore the relationship between the application's concurrency requirements and the capabilities of the architecture for implementing concurrency. Finding an efficient mapping between these two things is the core problem of the programmer's job. In the MESCAL project, we have developed a disciplined mapping methodology that allows programmers to understand and focus on this concurrency implementation gap. They are supported by efficient code generation to abstract from irregularly pipelined and forwarded datapaths described in Tipi.

Our goal is to create a deployment methodology that handles a wide variety of application models of computation and multiprocessor architectures. Here, we have covered the general high-level issues involved, and outlined the concepts of a general solution. In the next section, we will show how to apply these ideas to a specific application domain and a specific architecture – network processing and network processors.

2.   Deployment Case Study: Network Processing

In the preceding section, we introduced the general concept of a programming model and gave an overview of techniques applied to solve the compilation problem for general heterogeneous multiprocessor systems. In this section, we show how we simplify the deployment of programmable platforms for a certain application domain, namely packet processing. We derive an appropriate programming model that provides an efficient path to implementations. We also reveal techniques to automatically map threads of computation to processing resources. This automation is one major step for compiling applications to multiprocessor architectures. Our goal is to apply the lessons learned in this specialized domain to more general application and architecture descriptions based on Tipi.

In the following section, we introduce the programming model NP-Click for Intel's IXP1200 network processor. NP-Click is based on the Click [130] model of computation. It exposes certain architectural features to the programmer that are necessary to obtain good performance, while hiding less relevant architectural details. An example of an architectural feature visible in the NP-Click model is the partitioning of threads that are supported in hardware by the IXP microengines. Section 2.2 addresses how the programmer's burden can be reduced by automating the mapping of threads to microengines using integer linear programs (ILPs).

2.1   NP-Click: A Programming Model for the Intel IXP1200

The architectural diversity and complexity of network processor architectures motivate the need for a more natural abstraction of the underlying hardware. In this subsection, we describe a programming model, NP-Click, which makes it possible to write efficient code and improve application performance without having to understand all of the details of the target architecture. Using this programming model, we implement the data plane of two network processing applications, namely IPv4 forwarding and DiffServ, on a particular network processor, the Intel IXP1200. We compare the development process, achievable performance, and resource usage of the NP-Click based implementations with hand-coded Microengine C. Our results show that the applications written in NP-Click perform within 10% of hand-coded versions while reducing the development effort noticeably by up to a factor of four. The expense is increased resource requirements.

2.1.1   Introduction.   The past five years has witnessed over 30 attempts at programmable solutions for packet processing [218]. With these architectures, network processor designers have employed a large variety of hardware techniques to accelerate packet processing, including parallel processing, special-purpose hardware, heterogeneous memory architectures, on-chip communication mechanisms, and the use of peripherals, as discussed in

Chapter 3, Section 3. However, despite this architectural innovation, relatively little effort has been made to make these architectures easily programmable. In fact, these architectures are very difficult to program [131].

The current practice of programming network processors is to use assembly language or a subset of C. This low-level approach to programming places a large burden on the programmer to understand fine details of the architecture simply to implement a packet processing application, let alone optimize it. We believe the programmer should be able to implement an application using a natural interface, such as a domain-specific language. To accomplish this, we need an abstraction of the underlying hardware that exposes enough architectural detail to write efficient code for that platform, while hiding less essential architectural complexity. We call this abstraction a *programming model* (see Definition 6.1).

Our goal is to create an abstraction that enables the programmer to realize the full computational power of the underlying hardware. We realize that our programming model will initially introduce some implementation inefficiencies versus a hand-coded approach. However, our programming model offers a compelling design flow that produces implementations that are within 10% of the performance of a hand-coded approach (for a realistic packet mix) at a fraction of the design time.

This subsection describes NP-Click, a programming model for a common network processor, the Intel IXP1200. We illustrate our approach by using NP-Click to implement an IPv4 packet forwarder and a Differentiated Services interior node. The IPv4 packet forwarder is a performance focused benchmark, while the DiffServ application contains more functionality but supports fewer ports. We compare NP-Click with Microengine C [112] across three categories: Development process, achievable performance, and resource usage of the final implementation. We analyze these results and compare the advantages and disadvantages of the Microengine C and NP-Click programming environments.

The remainder of this section is organized as follows: The next subsection describes some background. Section 2.1.4 describes our programming model for the Intel IXP1200. We report our results in Section 2.1.5. Finally, we summarize our findings in Section 2.1.6.

2.1.2   Background.   In this subsection, we describe some relevant background to our work. We first give an overview of Click, a domain-specific language and infrastructure for developing networking applications, upon which our programming model is based. Next, we describe the Intel IXP1200, the target architecture for our application.

Click.   Click is a domain-specific language designed for describing networking applications [130]. It is based on a set of simple principles tailored for the networking community. Applications in Click are built by composing computational tasks, or *elements*, which correspond to common networking operations like classification, route table lookup, and header verification. Elements have input and output ports that define communication with other elements. Ports are connected via edges that represent packet flow between elements.

In Click, there are two types of communication between ports: *Push* and *pull*. Push communication is initiated by the source element and effectively models the arrival packets into the system. Pull communication is initiated by the sink and often models space available in hardware resources for egress packet flow. Click designs are often composed of paths of push elements and paths of pull elements. Push paths and pull paths connect through special elements that have different typed input and output ports. The Queue element, for example, has a push input but a pull output, while the Unqueue element has a pull input, but a push output.

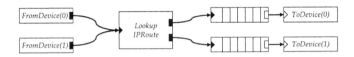

Figure 6.6.   Click diagram of a two-port packet forwarder.

Figure 6.6 shows a Click diagram of a simple two port packet forwarder. The boxes represent elements. The small triangles and rectangles within elements represent input and output ports, respectively. Filled ports are push ports, while empty ports are pull ports. The arrows between ports represent packet flow.

Click is implemented on Linux™ using C++ classes to define elements. Element communication is implemented with virtual function calls to neighboring elements. To execute a Click description, a task scheduler is synthesized to run all push (pull) paths by firing their sources (sinks), called schedulable elements.

A natural extension of this Click implementation is to multiprocessor architectures that may take advantage of the inherent parallelism in processing packet flows. A multi-threaded version of Click targets a Linux implementation and uses worklists to schedule computation [49]. Two pertinent conclusions can be drawn from this work: First, significant concurrency may be gleaned from Click designs in which the application designer has made no special effort to express it. Since packet streams are generally independent, ingress packets may be processed by separate threads with very little interaction. Second, a Click configuration may easily be altered to express additional concurrency without changing the application's functionality.

Intel IXP1200.   The IXP1200 [108] family is one of Intel's network processor product lines based on their Internet Exchange Architecture. It has six

RISC processors, called microengines, plus a StrongARM™ processor (see Figure 6.7). The microengines are geared for data plane processing and have hardware support for four threads that share a program memory. The StrongARM is mostly used to handle control and management plane operations. The memory architecture is divided into several regions: Large off-chip SDRAM, faster external SRAM, internal scratchpad, and local register files for each microengine. Each of these areas is under software control and there is no hardware support for caching data from slower memory into smaller faster memory (except for the small cache accessible only to the StrongARM™). The IX Bus (an Intel proprietary bus) is the main interface for receiving and transmitting data with external devices such as MACs and other IXP1200s. It is 64 bits wide and runs up to 104MHz allowing for a maximum throughput of 6.6Gbps. The microengines can directly interact with the IX bus through an IX Bus Unit, so a thread running on a microengine may receive or transmit data on any port without StrongARM™ intervention. This interaction is performed via Transmit and Receive FIFOs which are circular buffers that allow data transfers directly to/from SDRAM. For the microengines to interact with peripherals (e.g. to determine their state), they need to query or write to control status registers (CSRs). Accessing control status registers requires issuing commands across the command bus which is also used for issuing hash engine, scratchpad memory, and Transmit and Receive FIFO commands.

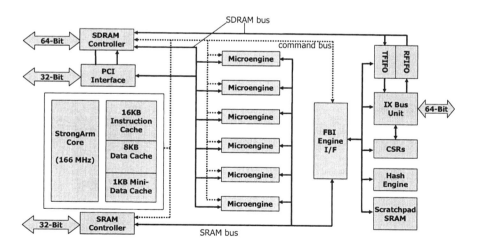

Figure 6.7. Intel IXP1200 architecture.

2.1.3   Programming Models.   There is currently a large gap between domain-specific languages that provide programmers a natural interface, like Click, and the complex programmable architectures used for implementation, like Intel's IXP1200. In this section, we introduce the concept of a programming model to assist in bridging this gap.

Implementation Gap.   We believe Click to be a natural environment for describing packet processing applications. Ideally, we would like to map applications described in Click directly to the Intel IXP1200. However, there is currently a large gap between Click and the low level programming interface the IXP1200 exposes. The simple yet powerful concept of push and pull communication between elements that communicate only via passing packets, coupled with Click's rich library of elements, provides a natural abstraction that aids designers in creating a functional description of their application. This is in stark contrast to the main concepts required to program the IXP1200. When implementing an application on this device, the programmer must carefully determine how to effectively partition the application across the six microengines, make use of special-purpose hardware, effectively arbitrate shared resources, and communicate with peripherals. We call this mismatch of concerns between the application model and target architecture the *implementation gap* (see Figure 6.8). To facilitate bridging this gap, we propose an intermediate layer, called a programming model (see Definition 6.1), which presents a powerful abstraction of the underlying architecture while still providing a natural way of describing applications.

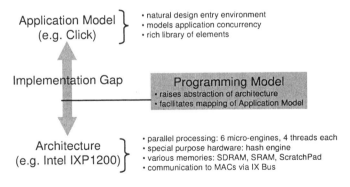

Figure 6.8.   Implementation gap.

Possible Approaches to Solving the Implementation Gap.   There are many different approaches to solving the implementation gap. We classify prior work in this area into four major areas:

- Library of application components.

- Programming language-based.

- Refinement from formal models of computation (MoCs).

- Run-time systems.

The *library of application components approach* exports a collection of manually designed blocks to the application programmer, who stitches these together to create the application. The advantage of such an approach is a better mapping to the underlying hardware, since the components are hand-coded. In addition, these components implement an abstraction that is natural for an application writer as the components are often similar to application model primitives. The disadvantage of this approach is the need to implement every element of the library by hand. If only a limited number of library elements are needed, this approach may be successful. However, in practice, we suspect a large number of elements are needed as application diversity grows. This problem is further compounded when a number of variants of each library element are needed [192]. For example, Intel has released a framework called ACE [107] that is a library-based approach built on assembly code. A library of Click elements written in C targeted at embedded processors is described in [208].

A *programming language approach* utilizes a programming language that can be compiled to the target architecture. With this approach, a compiler needs to be written only once for the target architecture. Then, all of the compiler's optimizations can be applied to all of the applications that are written for the architecture. The principal difficulty with this approach is the requirement to compile to heterogeneous architectures with multiple processors, special-purpose hardware, numerous task-specific memories, and various buses. The programming abstraction required to effectively create a compiler for such architectures would likely force the programming language to include many architectural concepts that would be unnatural for the application programmer. Examples of this alternative include the numerous projects that have altered the C programming language by exposing architectural features [112, 140, 168, 56].

Another class of approaches uses *refinement from formal models of computation (MoCs)* to implement applications. Models of computation define formal semantics for communication and concurrency. Examples of common MoCs include Kahn Process Networks [120] and synchronous dataflow [146]. Because they require applications to be described in a MoC, these approaches are able to prove properties of the application (such as maximum queue sizes required and static schedules that satisfy timing constraints). This class of solutions also emphasizes application modeling and simulation [17]. The disadvantage of this method is that implementation on heterogeneous architectures is inefficient because most implementation paths require significant compiler support. As an example, Edwards has written a compiler to implement designs described in Esterel, a language that implements the synchronous/reactive MoC [23]. However, his work generates C code and relies on a C compiler for implementation on target architectures [60]. In addition, the MoCs used by these approaches may not be natural design entry environments. For example POLIS requires all applications to be expressed in co-design finite state machines [17].

Run-time systems are another category of solutions to the implementation gap. Run-time systems introduce dynamic operation (e.g. thread scheduling) that enables additional freedom in implementation. Dynamic operation can also be used to present the programmer with an abstraction of the underlying architecture (e.g. an abstraction of infinite resources). While run-time systems are necessary for general-purpose computation, for many data-oriented embedded applications (like data plane packet processing) they introduce additional overhead at run-time. Additionally, some ASIP architectures have included hardware constructs to subsume simple run-time system tasks. Examples include thread scheduling on the IXP1200 and inter-processor communication (ring buffers on the Intel IXP2800 [111]). Examples of the run-time systems approach include VxWorks® [244] and the programming interface for the Broadcom Calisto™ [174].

Based on the trade-offs between the above approaches, we propose a programming model that is a hybrid of the application component library and programming language approaches. We believe that today's best practice of using a C-based language places undue burden on the programmer to generate even a functional implementation of an application, let alone an efficient one. It is this concern that motivates our own new layer to sit atop Microengine C. We describe our approach in the next section.

2.1.4   NP-Click: A Programming Model for the Intel IXP1200.   In this section, we describe NP-Click, our programming model as implemented on the Intel IXP1200. NP-Click combines an efficient abstraction of the target architecture with features of a domain-specific language for networking. The result is a natural abstraction that enables programmers to quickly write efficient code. The model is designed to ease three major difficulties of programming network processors: Taking advantage of hardware parallelism, arbitration of shared resources, and efficient data layout. This section describes the main components of the programming model: Elements and their communication, threading, and arbitration of shared resources. We also give hints for using NP-Click to arrive at an efficient implementation. A more detailed description can be found in [220, 221].

Overview of the Model.   Our programming model integrates concepts from Click to provide a natural design entry environment, and an abstraction of the IXP1200 architecture in order to leverage the computational power of the device.

To describe applications, we borrow Click's simple yet powerful abstraction of elements communicating by passing packets via push and pull semantics. Since our initial studies of the IXP1200 architecture showed the importance of multi-threading to hide memory and communication latency, we chose to export thread boundaries directly to the application programmer.

Unlike Click's implementation, elements in our programming model are implemented in Microengine C, the subset of C the IXP1200 supports. In addition, due to the performance impact of data layout on the target architecture (between registers, scratchpad, SRAM, and SDRAM), our implementation enables the programmer to effectively use these memories. Since the IXP1200 has separate program memories for each microengine, we allow multiple implementations of the same element of a design to exploit additional application concurrency. However, since most data memory is shared among microengines, the programmer must specify which data is shared among these instances and which data can be duplicated. We also provide the programmer with a machine API that hides pitfalls of the architecture and exports a more natural abstraction for unique memory features and co-processors. In addition, we provide a library that abstracts shared resources to separate the problem of arbitration from computation.

Elements.   Computation in our programming model is described in a fashion similar to Click, with modular blocks, called elements, which are sequential blocks of code that generally encapsulate particular packet processing functions. However, in our model, elements are defined using Microengine C, keywords for memory layout, and a machine API that provides key abstractions of low-level architectural details of the IXP1200.

Before describing the details of our programming model, it is important to understand the distinction between elements, types and instances. An element is a defined functional block within a design that has a type, which defines its functionality and the semantics of its ports. There may be multiple elements of the same type in a design. An instance is an implementation of an element. Depending on an application's mapping on to the target architecture, an element may have multiple instances to exploit parallelism.

Figure 6.9 shows a small Click network that illustrates the difference between a type, element, and instance. The boxes in the diagram represent elements. $FromDevice(0)$ and $FromDevice(1)$ are multiple elements of the same type. $LookupIPRoute$ is a single element with multiple instances (i.e. it is implemented by Thread zero and Thread one).

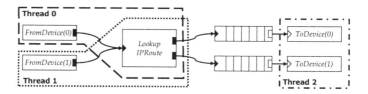

Figure 6.9. Example packet forwarder with thread boundaries.

Data Layout. As a significant portion of implementation speed is due to memory access latency, we provide some mechanisms when describing an element to guide memory layout.

To separate memory layout concerns from computation, we supply a facility to describe data by its scope. We provide four data descriptors:

- *Universal:* Data that is shared among all types.

- *Global:* Data that is shared among all elements of a specific type.

- *Regional:* Data that is shared among all instances of a specific element.

- *Local:* Data that is local to an instance.

The universal data descriptor describes data that needs to be accessible by all elements. Since this descriptor breaks the element abstraction, we aim to minimize the use of this descriptor. To date, we have not encountered applications that require this construct. We suspect it will mostly be used as an optimization.

Global data descriptors are used for data that must be shared across all elements of a given type. It could be used for a shared hardware resource that all elements of a particular type must use. For example, metering elements need to access a clock to determine the rate of packet flow.

Since elements in a Click design may be instantiated multiple times for performance reasons, the regional type modifier describes data within an element that must be shared across instantiations. For example, a *LookupIPRoute* element, which looks up the destination port of a packet, requires a large amount of storage for the routing table. As a result, to implement multiple threads that contain instances of the *LookupIPRoute* element without duplicating the entire routing table in memory, as shown in Figure 6.9, the lookup table must be shared among the instances but not among different *LookupIPRoute* elements.

The local data descriptor is used for state local to an element that need not be shared across multiple instantiations of an element. Examples of this type include temporary variables and loop counters.

Our abstraction is built on top of the *declspec* construct used in Micro-engine C to bind data to a particular memory (e.g. SRAM, SDRAM, scratchpad) at compile time. This may be used by the programmer for additional visibility into the memory architecture to improve performance of the implementation by specifying, for example, that certain large data structures, like routing tables, be placed in a large memory.

Machine API. In addition to data descriptors, our programming model hides some of the nuances of the Intel IXP1200 architecture. These abstractions are used in conjunction with Microengine C to describe computation within an element.

The IXP1200 implements special-purpose hardware for tasks that are commonly executed in software. To shield the programmer from the details of interacting with these hardware blocks, we export an application-level abstraction that encapsulates common uses of the hardware. For example, the IXP1200 has eight LIFO (last in, first out) registers that implement the common stack operations (push and pop) in a single atomic operation. However, these operations do not, for example, perform bounds checking or thread safety checks. We implement a lightweight memory management system that exposes a natural interface, namely $malloc()$ and $free()$ which makes use of the LIFO registers to implement a thread-safe freelist that performs bounds checking. These abstractions enable the programmer to reap the performance advantage of special-purpose hardware without understanding particulars of their implementation.

Communication. Our programming model borrows the communication abstraction from Click [130]: Elements communicate only by passing packets with push or pull semantics. However, our implementation of this abstraction is quite different from that of the Click software.

We define a common packet data layout that all elements use. We use a packet descriptor, allocated to SRAM, which stores the destination port and the size of the packet. The packet itself is stored in SDRAM. We define methods for reading and writing packet header fields and packet bodies, so these implementation details are hidden from the user.

As an optimization, we implement the packet communication by function calls that pass a pointer to the packet descriptor and not the packet itself. We enforce that compute elements do not send the same packet to multiple output ports to ensure that only one element is processing a particular packet at any given time. The packet data layout provides an abstraction that efficiently communicates packets among elements, but shields the programmer from the specifics of the IXP1200's memory architecture.

Threading. Arriving at the right allocation of elements to threads is another key aspect to achieving high performance. Thus, we enable the programmer to easily explore different mappings of elements to threads on the IXP1200. While we believe this task can be automated as shown in the following Section 2.2, given the great performance impact of properly utilizing threads, we make thread boundaries visible to the programmer.

As observed in [49], paths of push (pull) elements can be executed in the same thread by simply calling the source (sink). We implement a similar mechanism. However, because of the fixed number of threads on the IXP1200, we also allow the programmer to map multiple paths to a single thread. To implement this, we synthesize a scheduler that fires each path within that thread. For example, to implement the design in Figure 6.9, we would synthesize a round-robin scheduler for the schedulable elements in Thread 2 ($ToDevice(0)$ and $ToDevice(1)$). We hide the details of how to implement multiple schedulable elements within a thread from the user, but still give them the power to define thread boundaries at the element level.

Arbitration of Shared Resources.   The amount of parallelism present in the target architecture places pressure on shared resources. For example, the IXP1200 has 24 threads that may each simultaneously request a control status register. Such situations lead to potential hazards that necessitate arbitration schemes for sharing resources. To recognize the importance of sharing common resources, we separate arbitration schemes from computation and present them as interfaces to the elements. The two main resources that require arbitration on the IXP1200 are control status registers and the Transmit FIFO.

Control status registers are used to communicate with the MACs (e.g., determining which ingress ports have new data, which egress ports have space). Our experiments have shown access times to the control status registers ranging from nine to >200 clock cycles, with multiple simultaneous accesses sometimes leading to deadlock. The variability is due to contention on a common bus (the command bus) used for issuing SDRAM, SRAM, scratchpad, IX Bus, and control status register commands. This bus quickly saturates with multiple threads checking the status of the MAC at the same time. This variability is a critical factor in determining performance and one of the major difficulties of programming the IXP1200. To eliminate the need for the programmer to cope with this variability, we implement a per-microengine restriction on the number of concurrent control status register accesses. If a thread attempts to access a control status register while the maximum threshold of accesses are outstanding, a context swap is performed and another thread is loaded. While this may reduce overall microengine computational efficiency, this significantly reduces the variability in control status register access times. This abstraction wraps all reads and writes to the control status registers and is transparent to the programmer. This gives the user enough visibility to interact with peripherals efficiently while not allowing him to saturate the command bus.

The Transmit FIFO (TFIFO) is the principal mechanism for sending data off-chip and is shared by all threads. It is a 16-entry buffer used to queue data to be sent to the MAC. Each entry in the buffer has a flag that indicates whether that entry contains valid data. Microengines send data to the MAC by loading data into a TFIFO buffer entry and then setting the valid flag. A hardware state machine in the IX Bus Interface steps through the TFIFO as a circular buffer, waiting at each entry for the valid flag to be set. Once the valid flag is set for the current entry, the data in the entry is sent to the MAC and the state machine steps to the next entry. Microengines may query the valid flags and the entry the state machine is currently pointing to. Due to the unique nature of the TFIFO and the numerous threads that may be accessing it, managing the arbitration to this resource is a difficult problem. Perhaps the simplest arbitration scheme is to map a priori each port to one entry in the TFIFO. This eliminates the overhead associated with runtime coordination between elements but does not scale to applications with more than 16 ports. In addition, this scheme only performs well on an evenly distributed packet mix as the state machine will wait at a TFIFO entry even if there is no data to be sent on that port, thereby slowing the entire system. A slightly more complicated scheme that avoids these limitations is to allow threads to checkout any entry in the TFIFO. This requires a variable in shared memory to coordinate the free locations and mutex locking and unlocking. While this is able to scale functionally and can better handle a burst of data on a port, it imposes significant runtime overhead. To allow different arbitration schemes, we present an interface of the TFIFO to the elements. The implementation of this interface can be customized by the user without modifying the elements themselves.

Hints for Efficient Implementation. While NP-Click uses modularity to provide a productive method to implement networking applications, this incurs some performance overhead. In this section, we describe some hints for using NP-Click to achieve efficient implementations.

First, an element may exist in a thread which has multiple push paths or pull paths to service. To ensure the thread is making progress, elements should yield control when waiting on a long latency activity to complete or an intermittent event to occur. This is a coarser version of swapping threads on a multithreaded processor to hide memory access latency. For example, *ToDevice* often polls the MAC to determine whether to send more data. In this case, it is better for a thread if *ToDevice* checks the MAC once, then (if false) moves to another schedulable element. Whereas a multithreaded processor may implement thread swapping with multiple program counters and a partitioned memory space, swapping at the element level may be performed with static variables in an element instance.

While the programming model as presented may be sufficient for some applications, we concede it will fall short for others. As part of the path to final implementation, we provide facilities to further improve performance. These may include configuration-specific enhancements that might be encapsulated in a single element or optimizations across elements (like a specific scheduler for a set of schedulable elements). The enhancements we have used to date are minor changes to a single element or small modifications of arbitration schemes that greatly improve performance. By using NP-Click, the programmer is able to quickly pinpoint performance bottlenecks in the implementation.

Basis: Intel Microengine C.   Elements in NP-Click are implemented in Microengine C, the subset of C that the IXP1200 supports. NP-Click is thus an abstraction layer above Microengine C. This subsection describes characteristics of Microengine C for completeness.

The initial programming model that Intel provided for the microengines was assembly language [110]. It exposes all facets of the architecture under programmer control. Intel also provided a macro assembler that supports higher-level programming constructs like conditionals and loops. There is also a register allocator so symbolic variable names can be used.

Later, Intel augmented their assembly language interface to the microengines with a subset of C (Microengine C) [112]. Microengine C supports loops, conditionals, functions, intrinsics (function calls using C syntax that direct instruction selection), basic data types, and abstract data types, such as structs and bit-fields. However, data allocation to different memory regions is defined by the user. For practical applications, explicit binding is necessary at declaration time. In addition, the multithreading model is explicit. The programmer must manually divide their application across microengines and threads, control all inter-processor and thread communication, and arbitrate access to shared resources. Intel also provides a library that defines additional data types, macros, and functions that provide a slightly higher abstraction of the hardware. For example, there are bit-fields that export the format of control status registers and intrinsics for assembler instructions that use the hash engine.

2.1.5   Results.   We explore the effectiveness of our programming model by using it to describe the data plane of an IPv4 router and a DiffServ interior node on an Intel IXP1200. This section describes the application we implemented, the experimental setup used to gather data, and our results for maximum data rates for numerous packet mixes.

Applications.   For this case study, we use Microengine C and NP-Click to implement two applications: Packet forwarding and a DiffServ interior node. The IPv4 packet forwarding application is a performance-centric benchmark with relatively narrow functionality. The second application, a DiffServ interior node, is a functionally rich application with lower performance requirements.

IPv4 Packet Forwarding.   IP Version 4 packet forwarding [16] is a common kernel of many network processor applications. We chose to implement the data plane of a 16 port Fast Ethernet (16x100Mbps) IPv4 router. This application is based on the network processor benchmark specified in Chapter 2, Section 2. The major features of this benchmark are listed below:

- Incoming packets are checked for validity, including proper version number and correct header length.

- The egress port of a packet is determined by a longest prefix match route table lookup based on the IPv4 destination address field.

- After the egress port has been determined, the time-to-live (TTL) and checksum fields in the packet header are updated.

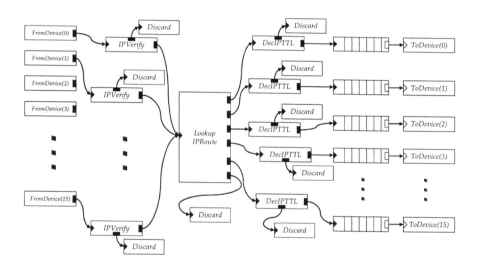

Figure 6.10.   Click representation of IPv4 data plane.

Figure 6.10 shows a graphical representation of the Click description of the router. We allocate 16 threads (four microengines) for receiving packets and eight threads (two microengines) for transmitting packets.

Differentiated Services Interior Node.   The differentiated services architecture (DiffServ) [24] is a method of facilitating end-to-end quality of service (QoS) over an existing IP network. In contrast to other QoS methodologies, it is a provisioned model, not a signaled one. This implies network resources are provisioned for broad categories of traffic instead of employing signaling mechanisms to temporarily reserve network resources per flow. A DiffServ network relies on traffic conditioning at the boundary nodes to simplify the job of the interior nodes. The boundary nodes of a DiffServ network aggregate ingress traffic into a number of categories, called *behavior aggregates* (BAs), using the *differentiated services codepoint* (DSCP) as specified in [173]. The interior nodes apply different *per hop behaviors* (PHBs) to each of the BAs. The classes of PHBs recommended by IETF include:

- *Expedited Forwarding (EF):* Low packet loss, latency, and jitter.

- *Assured Forwarding (AF):* Four subclasses of traffic, each with varying degrees of packet loss, latency, and jitter.

- *Best Effort (BE):* No guarantees of packet loss, latency, or jitter.

The PHBs in a DiffServ implementation are defined by a combination of:

- *Classifiers:* Elements that select a subset of the packet stream based on packet header fields.

- *Traffic conditioners:* Elements that measure, mark, shape, and drop packets.

For this case study, we implemented an interior DiffServ node. While there is less monitoring and shaping than in a boundary node, we believe an interior DiffServ node is a good benchmark for network processors as it is a functionally rich application that stresses different aspects of the development process.

Our DiffServ application begins with data-plane IPv4 packet forwarding functionality. After an ingress packet passes through IP verification, IP lookup, and time-to-live decrement, it is classified based on its DSCP, with each class of traffic receiving different treatment. For example, EF traffic is first metered, with traffic below the specified data rate queued for transmit, while traffic above this data rate is discarded. On egress, we use two cascaded packet schedulers:

- Deficit Round-Robin scheduling (DRR) [222] for EF and AF classes with a weighting toward EF and higher priority AF classes.

- Strict priority scheduling between the output of DRR scheduling and BE traffic.

The DiffServ application we implement supports 4 Fast Ethernet ports (4x 100Mbps). Though the IXP1200 may seem like overkill for this application, we attempted to implement an 8x100Mbps version of this application, but neither the Microengine C nor NP-Click implementation could support this line rate.

Setup. To test each implementation, we used a cycle-accurate architecture simulator of the IXP1200 [113] assuming a microengine clock rate of 200MHz and an IX Bus clock rate of 100MHz. Our simulation environment also models the Ethernet MACs (Intel IXF440s) connected to the IX Bus. For both applications, the destinations of the input packet streams were randomly distributed evenly across the output ports. In addition, the routing table contained 1000 entries and measurements were not taken until steady state was reached.

Comparison of our implementations in NP-Click to published results is difficult because relatively few are available for the Intel IXP1200. Of those results, little information is given about their experimental setup (e.g. IXP1200 and IX Bus clock speed, peripherals used, size of routing table, data rate measurement methodology). These details can have an enormous impact on the reported performance. Hence, for comparison to another implementation, we also hand-coded the applications in Microengine C. The forwarding application is based on the reference design supplied by Intel [109].

IPv4 Packet Forwarding. To measure performance for the IPv4 packet forwarding application, we consider the packet forwarder to be functional if it has a steady state transmit rate that is within 1% of the receive rate without dropping any packets. We tested the router with a 1000 entry routing table whose entries are chosen at random. The destinations of the input packet streams are randomly distributed evenly across output ports. We test each of the implementations with a variety of single packet size input streams (64, 128, 256, 512, 1024, 1280, and 1518 bytes) and the IETF Benchmarking Methodology Workgroup (BMWG) mix [32]. We use 64 and 1518 byte packet streams as they represent the minimum and maximum frame sizes permitted by the Ethernet standard. The packet sizes in between are included to give additional insight into the performance of the different implementations. The BMWG packet mix provides a more realistic input data set as it contains an even random distribution of seven packet sizes ranging from 64 bytes to 1518 bytes. For each input packet stream, we measure the maximum sustainable aggregate data rate.

Differentiated Services Interior Node. For the DiffServ application, measuring performance is not as simple since the specification requires non-conforming packets to be dropped. Thus, the transmit data rate will always be less than received data rate. To gauge performance we compare the egress data rates of the constituent traffic flows. For the baseline setup for all measurements, the ingress data rates were set to the following percentages of ingress bandwidth:

- Assured Forwarding, Class 1 (AFC1): 20%

- Assured Forwarding, Class 2 (AFC2): 15%

- Assured Forwarding, Class 3 (AFC3): 10%

- Assured Forwarding, Class 4 (AFC4): 5%

▪ Best Effort (BE): 10%

We measured egress data rates for all traffic flows as the Expedited Forwarding (EF) traffic grew from 0% to 40% of ingress bandwidth in 5% increments. When the EF flow is set to 40% of ingress bandwidth, the aggregate ingress bandwidth is at line rate (100Mbps/port). Since this is only a four-port design, we do not see the load balancing issues witnessed in the IPv4 packet forwarding implementations. Thus, to measure the worst case, the packet sizes of all flows were set to 64 bytes.

Performance.   This section describes the performance results of the Microengine C and NP-Click implementations. We first present the results for IPv4 packet forwarding, then for the DiffServ interior node.

IPv4 Packet Forwarding.   The results of our experiments for IPv4 packet forwarding are shown in Figure 6.11. The Microengine C implementation is able to perform at 85% of line rate (1360Mbps aggregate) across all single packet size input streams.   For the BMWG packet mix, the performance is slightly lower (1200Mbps aggregate) because of dynamic load balancing effects. We attribute the consistent data rate across all packet sizes to sub-optimal arbitration of multiple threads accessing the shared transmit FIFO as the performance limiting factor.

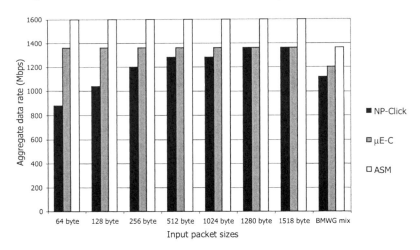

Figure 6.11.   Performance comparison of 16 port IPv4 packet forwarding implementations.

The aggregate bandwidth of the NP-Click implementation ranges from 880-1360Mbps.   NP-Click's implementation has more processing overhead per packet than Microengine C's. As a result, for data streams composed of smaller packets, NP-Click's throughput suffers. For the BMWG packet mix, a more realistic data set, the NP-Click version also suffers from load balancing issues, but is able to achieve 93% of the Microengine C implementation (1120Mbps aggregate).

For reference, we show the performance of a hand-tuned assembler implementation based on an Intel reference design. The assembler implementation was able to meet line rate (1600Mbps aggregate) for single packet size streams, but only maintains 1360Mbps when tested with the BMWG packet mix. Both the Microengine C and NP-Click implementations fall short of this, by 11.7% and 17.6% respectively, because of the fine degree of scheduling that is available only when programming at the assembly language level.

Differentiated Services Interior Node. Both the Microengine C and NP-Click implementations were able to receive packets at line rate. The egress data rates of all flows for both implementations are shown in Figure 6.12 and Figure 6.13. Ideally, the graph should be a horizontal line for Assured Forwarding (AFCx) classes indicating their egress data rates were not affected by increase in EF traffic. For EF traffic, the egress and ingress data rates should be equal through 20%, then egress EF traffic should level off. This is due to the setting of the bandwidth meter in the EF PHB. The egress Best Effort (BE) data rate is bound to decrease as EF increases since BE packets are subject to a strict priority scheduling with respect to all other flows.

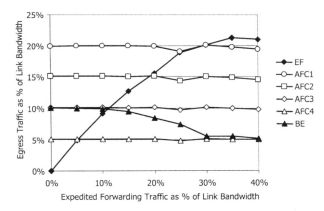

Figure 6.12. Microengine C DiffServ egress data rates.

For the Microengine C implementation, as the Expedited Forwarding (EF) flow increases, we see a minor decrease on the egress data rate of Assured Forwarding (AFCx) flows. This is caused by an overall increase in the ingress data rate, which results in more packets to process. This increases the total amount of computation required, which results in a slower packet processing rate. The decline in Best Effort (BE) egress bandwidth is caused by the increased availability of packets from other flows with higher priority.

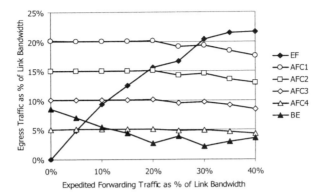

Figure 6.13. NP-Click DiffServ egress data rates.

The NP-Click implementation performs similarly, but experiences some additional degrading effects as the EF flow increases beyond 20% of ingress bandwidth. However, the egress bandwidths of EF and AFCx flows in NP-Click's implementation remain within 10% of Microengine C's. Since the transmit function is slightly slower in NP-Click, the BE data rate takes a performance hit earlier in the ramp up of EF traffic.

Development Process. This section compares and contrasts the development process of the applications using Microengine C and NP-Click. Specifically, we focus on the debugging and performance improvement process, design time allocation, and total design effort. Number of lines of code is often used as a proxy for design effort. However, when comparing vastly different programming methodologies, this metric can be very misleading. Instead, we measure person-hours.

Microengine C. When using Microengine C to implement IPv4 packet forwarding and the DiffServ interior node, most of the development effort was spent arriving at a functionally correct initial implementation. The remainder of the design effort was spent improving the implementation. More than half of that was fixing bugs that arose from thread interactions. Some of these interactions were not obvious from the design or the code, which made debugging even more difficult.

Given the relatively low level abstraction of Microengine C, both profiling and optimizing the implementation proved difficult. As a result, we were able to only implement and test a few design alternatives. All the design alternatives we implemented were incremental changes on the initial implementation. Large design changes to the implementation would have required even more design effort, with no guarantee of performance improvement.

Our total design effort for the IPv4 packet forwarder using Microengine C was 400 person-hours. For the DiffServ implementation, we started with a hand-coded data-plane implementation of an IPv4 packet forwarder, then added the DiffServ functionality. The total design effort for this implementation was 320 person-hours.

NP-Click. Using NP-Click, we began with a Click description of the application and we were able to create an initial functionally correct design within a few days. We spent the majority of the design effort exploring the design space of implementations: Pinpointing design bottlenecks, changing the mapping of elements to microengines, and simulating. Due to the modularity of NP-Click, profiling different implementations was easy. Performance improvement consisted of three major categories: Changing mappings of elements to threads/microengines, better implementations of elements, and lower overhead arbitration schemes. Relatively little effort was spent debugging and when an error was spotted, it was easy to pinpoint which portion of the implementation was the cause. Common errors included incorrectly specifying element configurations and minor functional bugs within an element.

Our initial implementation of the IPv4 packet forwarding application had low performance (480Mbps aggregate). The limiting factor of this implementation was a poor arbitration scheme of a shared resource (Transmit FIFO). The scheme used a shared variable among all *ToDevice* elements that requires a locking mechanism to access and update. As a result, the transmit threads spent the majority of their time attempting to acquire this lock. To alleviate this contention, we wrote a configuration-specific arbitration scheme that uses an a priori partitioning of the Transmit FIFO which obviates the need for the global variable to be shared across all *ToDevice* elements. Because of NP-Click's separation of computation and arbitration, this hand-coded optimization was simple to implement as it was fully contained within the Transmit FIFO library so no change to the *ToDevice* element was necessary. This arbitration scheme increased performance by 83-133%. The total design effort of this application was 100 person-hours.

For the DiffServ application, we were able to quickly arrive at an initial functional implementation. This initial implementation was a naïve mapping of elements to microengines, which we then optimized. The total design effort for the NP-Click four-port DiffServ implementation was 120 person-hours.

Resource Usage.   Resource usage of the final implementation is a key comparison metric for any embedded software development approach. For this case study, we compare the number of microengines used, MIPS required, and code size for the Microengine C and NP-Click implementations.

For the IPv4 packet forwarding application, both the Microengine C and NP-Click implementations used all six microengines on the IXP1200. For

DiffServ, Microengine C required only three microengines, while NP-Click required four. The additional microengine for the NP-Click implementation was needed because of instruction store limitations. The instruction store on the Intel IXP1200 is limited to 2048 instructions per microengine. It is important to note that this memory is not a cache, but a flat memory. As a result, all instructions to be executed on the microengines must fit in this space. We attempted to fit the NP-Click DiffServ application on fewer than four microengines, but often created functional partitions that exceeded the instruction store on a particular microengine. Thus, some of the NP-Click implementation effort was spent optimizing for code size, not performance. The modularity of NP-Click is responsible for the higher code size. We suspect better Microengine C compilation technology could significantly reduce this. A summary of the number of required microengines and code sizes for the four implementations is given in Table 6.1.

Table 6.1. Statistics of the final implementations.

	IPv4 Packet Forwarding		DiffServ	
	μengine C	NP-Click	μengine C	NP-Click
Number of microengines	6	6	3	4
MIPS	1196.6	1197.8	585.3	719.9
Code Size [instr]	3870	5591	2363	6090

The number of microengines used does not give a clear measure of the amount of computing power required for the different implementations. To measure this, we calculated the MIPS executed per implementation using a representative packet flow. On every clock cycle, a microengine on the IXP1200 is either: Executing an instruction, aborting an instruction (due to a mispredicted branch), stalled (waiting for multi-cycle access to return without context switching), or idle. We calculate MIPS by aggregating the number of cycles spent executing and aborting instructions across all microengines and dividing by time. Since the microengines are running at 200Mhz, the peak MIPS rate is 200 per microengine. Table 6.1 includes a summary of our results. For IPv4 packet forwarding, the Microengine C and NP-Click implementations executed at similar MIPS rates. We attribute the extra MIPS in the NP-Click DiffServ implementation to overhead introduced by NP-Click's modularity and time spent in polling loops.

Discussion. In this case study, we have compared two different software development approaches for the Intel IXP1200, a common network processor. We compared Microengine C and NP-Click by implementing a 16 port IPv4 packet forwarder and a four-port DiffServ application. The advantages and disadvantages of each approach are discussed in the following.

Microengine C. The principal advantage of the Microengine C software development approach is resource usage of the final implementation. For the DiffServ implementation, we were able to meet line rate using only three microengines and used significantly fewer instructions. Microengine C also has a slight performance edge over NP-Click: It supports 75Mbps per port (BMWG packet mix) for the IPv4 packet forwarder and has a slightly higher egress data rate across all traffic flows for the DiffServ application.

Using Microengine C for software development had some drawbacks as well. The overall development effort per application was much longer than NP-Click's, as was the time to reach functional correctness of the application. After meeting the functional requirements, performance tuning was also quite difficult. Hence, all performance improvements were incremental changes on the initial implementation. As a result, with Microengine C, the final implementation is largely dependent on the initial implementation. This places a large burden on the programmer's intuition of how the application should be partitioned.

NP-Click. NP-Click's primary advantage is ease of programming. With NP-Click, we were able to rapidly implement applications to meet the functional specification. Then, we were able to easily explore the design space of implementations to further improve performance. As a result, the total design effort was 2.5-4x shorter than Microengine C's. We believe that this flow illustrates a typical usage of the programming model: First, NP-Click will be used to quickly gain functional correctness. Performance bottlenecks in the NP-Click implementation are then identified, and where needed, hand-coded optimizations will be applied. In our experience, these optimizations are easy to implement and localized to an element or a thread. Moreover, NP-Click produces code that is easier to debug, improve, maintain, and reuse.

For the IPv4 packet forwarder, the NP-Click implementation was able to route packets at 70Mbps per port (BMWG packet mix), 6.7% less than the Microengine C implementation. For DiffServ, the egress data rate of higher priority flows in NP-Click's implementation was within 10% of the Microengine C implementation.

The major weakness of using NP-Click is resource usage. For the DiffServ implementation, NP-Click was able to receive packets at line rate, but required an additional microengine. This was mainly due to NP-Click's code size overhead when compared to Microengine C. The instruction store limitations on the IXP1200 forced us to use one more microengine.

It is important to note that we spent significantly less effort using NP-Click for both applications. We believe further effort using NP-Click will result in better implementations. For the 16 port packet forwarding design, we believe

NP-Click can produce an implementation that can meet a higher data rate. For the DiffServ implementation, we can likely achieve similar performance with lower resource usage (but not as low as Microengine C's).

2.1.6   Conclusions.   As application complexity increases, the current practice of programming network processors in assembly language or a subset of C will not scale. Ideally, we would like to program network processors with a network application model, like Click. However, the implementation gap between Click and network processor architectures makes efficient implementation difficult. In this subsection, we defined NP-Click, a programming model that bridges this gap by coupling the natural expressiveness of Click with an abstraction of the target architecture. This enables efficient implementation. We have evaluated the usefulness of NP-Click for two network processing applications on the Intel IXP 1200 processor. In order to quantify the trade-off among performance, development effort, and resource usage we compared our NP-Click based implementations with hand-coded Microengine C.

The results can be summarized as follows:

- If performance is the primary goal, Microengine C has a slight edge, at the cost of design effort.

- If resource usage is predicted to be tight, Microengine C is the preferred approach.

- For effort-bound projects, NP-Click is preferred. NP-Click gives designers a fast path to an initial implementation and the facilities to try many different implementations to improve performance and resource usage. In either case, if there is little intuition about an ideal functional partitioning *a priori*, NP-Click is more attractive.

It may be possible to combine the two software development approaches: Use NP-Click to quickly try many different functional partitionings, yet write the final implementation in Microengine C. This would provide a compromise between design effort, performance and resource usage.

The Intel IXP1200 shares many of the salient features with other network processors: Multiple multithreaded processors, disparate memories with varying latencies, numerous heterogeneous shared resources, and different on-chip communication mechanisms. As result, we believe these results are applicable to many other network processors, including newer architectures.

With NP-Click, one major issue in achieving high performance that remains in the hands of the designer is finding a good partitioning and mapping of threads onto microengines. For the IXP1200 with its six microengines, the optimization of the mapping might still be feasible by hand. However, for next generation platforms like the IXP 2800 with its 16 microengines and eight threads of execution per engine it will become impossible to do this task manually. In the following section, we introduce a problem formulation that allows us to find an optimal mapping automatically.

2.2 Automated Task Allocation on Single Chip, Hardware Multithreaded, Multiprocessor Systems

The mapping of application functionality onto multiple multithreaded processing elements of a high performance embedded system is currently a slow and arduous task for application developers. Previous attempts at automation have either ignored hardware support for multithreading and focused on scheduling, or have overlooked the architectural peculiarities of these systems. This work attempts to fill the void by formulating and solving the mapping problem for these architectures. In particular, the task allocation problem for a popular multithreaded, multiprocessor embedded system, the Intel IXP1200 network processor, is encoded into a 0-1 Integer Linear Programming problem. This method proves to be computationally efficient and produces results that are within 5% of aggregate egress bandwidths achieved by hand-tuned implementations on two representative applications: IPv4 Forwarding and Differentiated Services.

2.2.1 Introduction. For a number of years, computer architects have tried many techniques to improve the performance of application-specific programmable processors. One of the most effective is multithreading with hardware support, of which there are many flavors including fine-grain multithreading, coarse-grained multithreading [96], and simultaneous multithreading [234]. Each approach attempts to maintain high processor utilization by having hardware dynamically schedule multiple threads. Growing silicon capability has now made it possible to incorporate multiple multithreaded processors on a single die. However, since programming environments for these architectures have not advanced as rapidly, software developers face the daunting challenge of efficiently allocating the tasks of their application onto hardware multithreaded, multiprocessor architectures. While some automated solutions exist for solving this problem generically (e.g. dynamic partitioning in operating systems, offline scheduling algorithms), for high performance embedded systems the most common solution is simply a manual partitioning of the design across threads and processors. Partitioning may be done either informally, by the designer's intuition and experience, or more methodically, by architecting the design to be flexible, and then changing the task allocation based on profiling feedback. In either case it is a time-consuming and challenging problem, largely due to a huge and irregular design space further exacerbated by resource constraints.

This work automates the task allocation process for hardware multithreaded, multiprocessor architectures. We use simplified models of the application and the architecture and leverage recent advances in 0-1 *integer linear programming* (ILP) to solve it efficiently. To demonstrate our approach, we map the data plane of an IPv4 router and a Differentiated Services interior node onto the Intel IXP1200, a hardware multithreaded, multiprocessor designed for network applications. For both examples, the runtime of our approach is less than one second, with resulting implementations performing within 5% aggregate data rate of hand partitioned designs.

The remainder of this section is organized as follows: The next subsection gives some background on prior work. The problem formulation is described in Section 2.2.3. In Section 2.2.4, we present the results of our approach for two network applications. Section 2.2.5 concludes and summarizes this work.

2.2.2   Related Work.

Mapping Problem.   The mapping of application tasks onto an embedded multiprocessor architecture is typically conducted in two steps: Task allocation and scheduling. Approximation algorithms have been extensively studied to solve these problems for general multiprocessor models [48, 217]. However, such generalized solutions are not suitable for modern embedded architectures since they fail to consider practical resource constraints. In particular, the approximation schemes do not take into account thread and storage limitations, which are critical factors that affect the quality of the mapping to multithreaded architectures. This simplification substantially limits the design space that would be explored for some embedded systems. Therefore, existing approximation algorithms are not appropriate to solve the task allocation and scheduling problems for hardware multithreaded multiprocessors.

ILP-Based Solution.   We utilize the framework of ILP to solve our variant of the mapping problem. The use of ILP in high-level synthesis is not new. Hwang, et al. [103] presented an ILP model for resource-constrained scheduling and developed techniques to reduce the complexity of the constraint system. Extending this concept, *mixed integer linear programming* (MILP) based task allocation schemes for heterogeneous multiprocessor platforms have been advanced [20]. These formulations determine a mapping of application tasks to hardware resources that optimizes a trade-off function between execution time, processor, and communication costs. The advantage of ILP is the natural flexibility to express diverse constraints and its potential to compute optimal solutions with reference to the problem model. However, ILP approaches are typically impractical, since most ILP solvers suffer from large run times even for simple problem instances. To counter this issue, we use a modern 0-1 ILP solver with improved search heuristics and additionally introduce special constraints to restrict the search space. Thus we exploit the flexible framework of ILP to generate optimal solutions to the mapping problem within fractions of a second.

The work that comes closest to our problem of mapping to hardware multithreaded architectures was published by Srinivasan et al. [225]. The authors consider the scheduling problem for the Intel IXP1200 and present a theoretical framework in order to provide service guarantees to applications. However, they do not consider practical resource constraints of the target architecture, nor do they test their methodology with real network applications. In contrast, our approach provides an efficient solution to the mapping problem, explicitly taking into account resource constraints of the hardware multithreaded multiprocessor.

Implementation Environment. The IXP1200 [108] is one of Intel's first network processors based on their Internet Exchange Architecture and has already been discussed in Subsection 2.1.2. There are several programming environments that may be used to design applications for the IXP1200 including NP-Click (see the preceding Section 2.1), Intel's ACE framework [107], and Intel's Microengine C [112]. We choose to use NP-Click as our implementation environment due to its modularity and ease of use. It is an efficient programming approach that provides visibility into salient architectural details that greatly affect performance.

2.2.3 Problem Formulation.
We approach the mapping problem in three steps:

- Construct a simplified model to capture only those salient application parameters and resource constraints that are most likely to influence the quality of the final solution.

- Encode the constraint system as a 0-1 ILP formulation.

- Solve the optimization problem using an efficient solver to determine an optimal, feasible configuration.

We elaborate on these steps in the following sections.

Model and Metric. We view the multiprocessor as a symmetric shared memory architecture consisting of a memory hierarchy in which each region has uniform access time from any processing element (PE). This obviates the necessity to explicitly account for memory and communication metrics in the model. We incorporate instruction store limits per PE as a resource constraint. In our experience, applications implemented on the IXP1200 are often instruction store bound, severely complicating the load balancing process. Furthermore, the tradeoff between instruction store and execution cycles of individual tasks is critical to determining optimal performance.

In our application model, tasks are classified by *class*. Elements within a *class* are functionally the same. Tasks within a *class* can have different *implementations*, which may differ in their execution cycles, number of instructions, and so on. In a multithreaded system, two tasks on the same PE may share part of the instruction store if they are of the same class and implementation. Such instructions are called *shareable instructions*. Conversely, a task that contains state may have an implementation in which instructions directly address specific state variables. Therefore these instructions may not be shared. However, since some multithreaded architectures allow for context relative addressing, they enable a logical separation of a single memory resource. As a result, instructions which utilize context relative addressing to reference state variables directly may be shared across threads, but not within one. Such instructions are called *quasi-shareable*. While direct references are faster, a designer may implement a task with state with *shareable* instead of *quasi-shareable* instructions by indirectly addressing state variables. Tasks written with shareable instructions incur additional cost of execution time and total instruction store, but allow a developer to tradeoff execution cycles and total instruction memory.

Further, we assume that the application consists of independently executing tasks. The queues in the data plane of a typical network application decouple tasks from execution dependencies. In other words, while there is a graph that represents dataflow through the application, the individual tasks may be considered as executing independently. Currently, we assume tasks have the same periodicity, but our formulation could be easily extended to accommodate multiple execution rates. The utilization of a PE by a task is measured by the number of execution cycles it consumes (execution time less long latency events). Our goal is to allocate tasks onto PEs with the objective of minimizing the average *makespan* – the maximum execution cycles of all tasks running on the system. We acknowledge that these assumptions sacrifice some accuracy, but they were carefully chosen to maximize the exploration of critical parts of the design space while still keeping the problem tractable.

Problem Formulation.   We attempt to solve the following resource-constrained optimization problem: Given a set of independent tasks and a set of PEs, find a feasible implementation for each task and a mapping of tasks to PEs so that the *makespan* is minimized.

Formally, we have a collection of task classes Y, and each class $y \in Y$ consists of a set of tasks $T(y) = \{t_{y_1}, t_{y_2}, ...\}$ and a set of implementations $M(y) = \{m_{y_1}, m_{y_2}, ...\}$. An individual task in $T(y)$ can operate in any one implementation from $M(y)$. Each $m_{y_i} \in M(y)$ is characterized by a tuple,

$(e_{y,m_{y_i}}, s_{y,m_{y_i}}, q_{y,m_{y_i}})$ denoting the number of *execution cycles*, the number of *shareable instructions*, and the number of *quasi-shareable instructions*, respectively.

- All tasks of the same class and implementation may share a part of the program instructions in a PE denoted by $s_{y,m_{y_i}}$.

- The quasi-shareable instructions denoted by $q_{y,m_{y_i}}$ can only be shared by at most N tasks of the same class and implementation, where N is the number of hardware threads in a PE.

- Once more than N tasks are assigned to a PE, extra instructions must be used to accommodate additional state registers.

By choosing the appropriate implementation for each task, the optimization problem attempts to minimize the *makespan*, while satisfying constraints on the instruction store. The set P consists of the available PEs in our system. An instruction store limit S_{limit} is enforced for each PE. The parameter E_{cycle} denotes the bound on *makespan*.

We encode the resource constrained decision problem as a 0-1 ILP; the variables in our constraint system are the following:

- (1) $x_{y,t,m,p}$: A 0-1 variable which indicates whether task $t \in T(y)$ belonging to task class $y \in Y$ with implementation $m \in M(y)$ is assigned to PE $p \in P$. In a feasible configuration, a variable with value one denotes a selection of implementation and PE for each task.

- (2) $a_{y,m,p,k}$: A 0-1 variable which indicates whether k or more tasks from class $y \in Y$ with implementation $m \in M(y)$ are assigned to PE $p \in P$, where $k \in \{1, 1 + N, 1 + 2N, \cdots, 1 + N \cdot \lceil (|T(y)| - N)/N \rceil\}$. This counts the number of tasks from class y and implementation m in PE p in increments of N.

The constraints in our system are the following:

$$\sum_{p \in P} \sum_{m \in M(y)} x_{y,t,m,p} = 1 \quad \forall y \in Y, \ \forall t \in T(y) \tag{6.1}$$

$$\sum_{y \in Y} \sum_{t \in T(y)} \sum_{m \in M(y)} e_{y,m} \cdot x_{y,t,m,p} \le E_{cycle} \quad \forall p \in P \tag{6.2}$$

$$a_{y,m,p,k} = 1 \Leftrightarrow \sum_{t \in T(y)} x_{y,t,m,p} \ge k \quad \forall p \in P, \ \forall y \in Y, \tag{6.3}$$

$$\forall m \in M\left(y\right),\ \forall k \in \left\{1, 1+N, 1+2N, \cdots, 1+N \cdot \left\lceil \left(\left|T\left(y\right)\right| - N\right)/N\right\rceil\right\}$$

$$\sum_{y \in Y} \sum_{m \in M(y)} \sum_{k \in \{1, \cdots, 1+N \cdot \lceil (|T(y)|-N)/N\rceil\}} q_{y,m} \cdot a_{y,m,p,k} +$$

$$\sum_{y \in Y} \sum_{m \in M(y)} s_{y,m} \cdot a_{y,m,p,1} \le S_{limit} \qquad \forall p \in P \qquad (6.4)$$

Constraint (6.1) is the exclusionary constraint that specifies that each task must be executed in exactly one implementation and assigned to exactly one PE. The total execution time of all tasks in their selected implementations in each PE must be less than E_{cycle} and this is ensured by constraint (6.2). Constraint (6.3) determines the values of the $a_{y,m,p,k}$ variables, which are then used in (6.4) to stipulate a bound on the combined instruction store of all tasks assigned to a PE, accounting separately for the shareable and quasi-shareable parts. The total number of variables in our constraint system is of order $O\left(max\left\{\left|T(y)\right| \mid y \in Y\right\} \cdot max\left\{\left|M(y)\right| \mid y \in Y\right\} \cdot \left|P\right| \cdot \left|Y\right|\right)$. The number of constraints is linear in the number of variables.

Solver. The search strategy is to perform a binary search on E_{cycle} to find the optimum possible execution time and a corresponding implementation and PE assignment for each task. We note that the decision problem formulated above is a reduction of the basic bin-packing problem and hence is NP-complete. Though encoding the problem as ILP allows us the flexibility to specify varied constraints, solving such problems in the general case is inefficient for reasonably sized instances. However, we take advantage of recent advancements in search algorithms and heuristics for solving 0-1 ILP formulations to efficiently compute solutions that are optimal with respect to our problem model. We use GALENA [43], a fast pseudo-Boolean SAT solver, to solve the constraint system.

Additionally, we can introduce specialized constraints to prune the solution space and remarkably speed up the ILP search procedure. For instance, tasks from the same task class and implementation are identical (since they are characterized by the same number of instructions and execution cycles), and hence lead to symmetric configurations. Accordingly, introducing symmetry-breaking constraints eliminates all redundant configurations that are distinguished only by a permutation of tasks from the same class. To generate these constraints, a total ordering is introduced over all tasks $(T(y), \le)$ and implementations $(M(y), \le)$ in each task class $y \in Y$ and PEs $(P, \le)$ in the system. Intuitively, let $\rho\left(t_{y_i}\right)$ and $\omega(t_{y_i})$ denote a selection of PE and implementation, respectively, for some task $t_{y_i} \in T(y)$. In order to break symmetry, we enforce the constraint that if t_{y_i} and t_{y_j} are two tasks in the same task class $y \in Y$ with

$t_{y_i} \leq t_{y_j}$, then

$$\rho(t_{y_i}) \leq \rho(t_{y_j}) \quad \wedge \quad \rho(t_{y_i}) = \rho(t_{y_j}) \implies \omega(t_{y_i}) \leq \omega(t_{y_j}).$$

In this way, we avoid symmetric configurations that arise due to a permutation of tasks. This directs the ILP search towards useful configurations and achieves significant run time speedups.

Similarly, we can exploit symmetries that arise because the PEs in the system are non-distinct. We may also include restrictions on the placement or pairing of certain tasks to expedite the search. The ILP framework is sufficiently general to accommodate various other user-specified constraints based on knowledge of the problem instance. Thus, the combination of a modern 0-1 ILP solver with specialized constraints to restrict the search space provides a powerful mechanism to solve the task allocation problem efficiently.

2.2.4   Results.   To demonstrate the validity of the model, we used the IXP1200 and two common and representative network applications, packet forwarding in IPv4 and a Differentiated Services (DiffServ) interior node. These applications with their configurations have already been introduced in the context of NP-Click in Subsection 2.1.5.

Table 6.2.   IPv4 Forwarding task characteristics.

Class	Number of Tasks	Execution Cycles	Shareable	Quasi-shareable
Receive	16	337	801	0
Transmit (Impl 1)	16	160	348	0
Transmit (Impl 2)	16	140	5	285

In both applications, a developer supplied the tasks to be used, and each class and implementation was profiled in a cycle-accurate simulation environment. To obtain average execution cycles per task, the application was tested with worst-case input traffic. An instance of each task class and implementation was run on a PE by itself in a functionally correct configuration so that it could be profiled with the appropriate traffic. We note that different configurations may cause a task to perform differently, but those effects have not yet been substantial. We have seen at most a 10% change in execution cycles consumed between a task compiled alone and in the presence of other tasks. For shareable and quasi-shareable instructions, the application was complied with varying task configurations. We implemented a 16 port Fast Ethernet (16x100Mbps) IPv4 router consisting of 16 Receive class and 16 Transmit class tasks with characteristics shown in Table 6.2. A Transmit task has state and can be written with either shareable or quasi-shareable instructions to reference its shared variables. For this application, we targeted a version of the IXP1200 for which the bound on the instruction store was 1024 instructions.

The result is optimal with respect to the model: The instruction store constraints preclude the two task classes from coexisting on any PE. The problem then degenerates into bin packing, the only wrinkle being that Transmit class tasks may exist in either implementation. Implementation two is preferred for all Transmit tasks since it is faster of the two and fits within instruction store limits. The resulting configuration shown in Table 6.4 is exactly the same partition that was arrived at from hand tuning.

Table 6.3.   DiffServ task characteristics.

Class	Number of Tasks	Execution Cycles	Shareable Instructions
Receive	4	99	462
Lookup	4	134	218
DSBlock	4	320	1800
Transmit	4	296	985

Our DiffServ application supported four Fast Ethernet ports (4x100Mbps) targeting the 2K instruction store version of the IXP1200. The corresponding task configurations are presented in Table 6.3. Since there are only four tasks in each class and the IXP1200 has four hardware threads per PE, the quasi-shareable instructions are omitted and all instruction store used is represented by the sharable component. After testing with various mixes of traffic flows, we found the egress bandwidth of each packet class in the automatically generated design to be within 2% of the hand-tuned design, except for Best Effort which occasionally transmitted at only one-third the data-rate of the hand-tuned. This is because there is a strict priority scheduler between Best Effort traffic and all other traffic, and that the latency of Transmit tasks is higher due to the simplifications inherent to the model. When the Transmit tasks service the various flows and fail to keep up with ingress, Best Effort is the first to suffer. Overall the automated partition is within 5% of the hand-tuned aggregate bandwidth for all data points, and was generated in less than a second while the hand-tuned design took days to arrive at.

The principal reason the hand-tuned design performs better than the automatically generated one is that the designer's internal model accounts for more aspects of the mapping problem than the one proposed here. Empirically, the major difference between these two models can be accounted for by the consideration of execution cycles consumed by polling. In our setup, tasks poll to determine whether a packet is ready to be processed, so there are execution

Table 6.4.   Final partitioning.

Application	PE1	PE2	PE3	PE4	PE5	PE6
IPv4 forwarding (hand & auto)	4 Receive	4 Receive	4 Receive	4 Receive	8 Transmit (Impl 2)	8 Transmit (Impl 2)
DiffServ (hand)	4 Receive	4 Lookup	2 DSBlock	2 DSBlock	2 Transmit	2 Transmit
DiffServ (auto)	1 Receive 1 Lookup 1 Transmit	1 Receive 1 Lookup 1 Transmit	2 DSBlock	2 DSBlock	1 Receive 1 Lookup 1 Transmit	1 Receive 1 Lookup 1 Transmit

cycles consumed by these polling loops. The execution cycles shown in Table 6.2 and Table 6.3 do not include this polling since it depends greatly on the final configuration, however it can impact design decisions. Consider the DiffServ allocation problem once the four DSBlocks are assigned to two PEs by themselves (a decision that both approaches agree on). We are left with a sub-problem of allocating four Receive, four Lookup, and four Transmit tasks to the four remaining PEs. Based on the model presented, since there is ample instruction store in the sub-problem, the optimal result is to put one of each task class into each PE. But since the tasks with shorter execution times (Receive and Lookup) spend execution cycles in their polling loops, a significant number of additional execution cycles are consumed: A fact which is not considered by the model. A knowledgeable developer would note that pairing an execution-heavy Transmit task with only one other execution-heavy task (one additional polling loop) might be better than putting it with the two execution-light tasks (two additional polling loops). In this new configuration, the Transmit tasks are paired together and Receive and Lookup are given their own PEs. After trying the configuration, the results indicate that the two polling loops Receive and Lookup introduce mitigate any apparent savings from being execution-light.

2.2.5   Conclusions.   As the functionality expected from single chip, high performance, multiprocessor systems continues to increase, the task of distributing that functionality becomes more critical. Designers are already faced with a large and non-intuitive set of tradeoffs for task partitioning and mapping. To cope with this, we formulate the mapping problem for one instance of such resource constrained embedded systems. By encoding the mapping problem as a 0-1 ILP, we allow for flexibility and extensibility while still utilizing a high performance back end. Problems are solved in less than a second with results that are optimal with respect to the model. The model has proved itself for two representative network application producing results within 5% aggregate bandwidth of hand balanced designs. While more applications need to be tested, we feel that this approach is robust and fast enough to be used as a tool by designers when developing software for these systems. It is one piece in an overall design process that will enable designers to explore different task implementations and identify optimal mappings and complements the NP-Click programming model introduced in the preceding section. This in turn expedites the overall application design flow for multiprocessor, hardware multithreaded embedded systems.

We aim to extend and improve this approach in a number of ways. First, we plan to address some limitations discussed earlier and incorporate effects like contention and excessive polling. Second, to further automate the design process, we plan to look at generating efficient task configurations from high-level application descriptions that will result in better designs. We also intend to test this approach on other more complex architectures.

3.   Conclusion

There are several prerequisites for a successful deployment of a programmable platform. In this chapter, we have focused on the issue of how to provide a design framework that, apart from design entry and evaluation, is able to ease the deployment of ASIPs and programmable platforms. Clearly, on the lowest level of abstraction, we expect the design tool to generate an assembler and a simulator of the described programmable architecture. The abstraction level is often increased by using a high-level programming language compiler that may also be generated from the architecture description. Still, the platform may be difficult to exploit since common programming languages cannot express concurrency and lack efficient access to special-purpose hardware. We therefore have introduced the notion of a *programming model*, which is a more general abstraction of the underlying hardware for the application programmer. Ideally, it provides enough visibility of hardware features and enough opacity from implementation details so that the application developer can focus on the actual behavior of the application. The challenge is now to efficiently translate (compile) from a programming model down to the platform.

The general solution of this translation is beyond the scope of this book. We have thus outlined the support for successful programmable platform deployment in Tipi and given a short overview of the compilation techniques pursued in the MESCAL project. As an illustrative example of how a programming model can be defined dedicated to a certain application domain, we have pointed out a Click-based programming model that is natural in the domain of network processing. Based on this model we have revealed how the compilation to a multi-processor core can be automated by using an ILP-based problem description.

In conclusion, we believe that better programming abstractions exist than contemporary programming languages, and that there is an efficient path from these abstractions to high-performance implementations on the underlying hardware.

II

PART II: USING COMMERCIAL TOOLS TO APPLY THE MESCAL METHODOLOGY FOR BUILDING ASIPS

Chapter 7

DESIGNING AND MODELING MPSOC PROCESSORS AND COMMUNICATION ARCHITECTURES

Heinrich Meyr[1,2], Oliver Schliebusch[2], Andreas Wieferink[2], David Kammler[2], Ernst Martin Witte[2], Olaf Lüthje[2], Manuel Hohenauer[2], Gunnar Braun[2], Anupam Chattopadhyay[2]

[1] *CoWare, Inc.*
San Jose, CA

[2] *RWTH Aachen University*
Institute for Integrated Signal Processing Systems
Aachen, Germany

In this chapter, the mapping of the MESCAL methodology onto the CoWare tool suite is discussed. In order to understand the mapping it is helpful to first understand the key ideas and properties of the tools. First of all, the CoWare tools address both issues of designing and programming a platform: The design of the computational elements as well as their interconnects.

The embedded processor designer (LISATek[TM]) allows the user to jointly design the architecture of an ASIP and the software development tools. In LISATek the ASIP is captured by a *single* model from which the tools (C-compiler, assembler, linker, instruction-set simulator, etc.) and the RTL hardware implementation are generated automatically, thus allowing an iterative design process to evaluate architectural alternatives. The early consideration of C-compiler requirements to the architecture guarantees the successful deployment of the architecture.

LISATek offers the designer the full architectural design space. The possibilities range from designing a simple computational element with only a very few instructions to using advanced architectural concepts such as VLIW, SIMD, etc. in a complex processor. Thus, the designer has the possibility to optimally match the architecture to the requirements. From an economical standpoint designing an in-house processor is a very attractive proposition since the IP is fully protected and costly license fees and royalties are eliminated.

The tool suite can be viewed as a workbench which separates the automatic generation of tools from optimization tasks. For example, in the implementation path the designer has the option to employ various optimization techniques on top of logic synthesis. Another example is instruction set synthesis. This open architecture allows the designer to incorporate innovative optimization strategies when they become available. This is of key importance as the optimization of ASIPs is a very active area of research.

The interconnect of the processing elements (ASIP) of a platform is as important as the elements themselves. As the traditional bus structures do not meet future on-chip communication requirements, the trend is towards fully network-on-chip (NoC) concepts. A tool suite must be capable to model these networks at various abstraction levels, both temporal and spatial. The philosophy underlying the CoWare tool suite (ConvergenSC^{TM}) employs the principle of orthogonalization of concerns [126]. Analogously as in telecommunication networks, the communication services are offered independently of the actual network topology. This allows the designer of a platform to explore network alternatives. For example, it allows the designer to trade computation versus communication ("the network is the computer").

So far we have discussed the two tool boxes to design and program platforms individually. As has been emphasized throughout in the introduction of this book the two elements must work seamlessly together in order to be useful in practice (see element 5: The successful deployment). This is of utmost importance for the actual user of the platform: The application programmer. He needs access to a virtual platform long before silicon is out, as discussed later on in the chapter.

1. Benchmarking the Application Requirements

Element 1 – Judiciously Using Benchmarking: *"ASIP development by definition must be application driven. [...] A benchmark must consist of a functional specification, a requirements specification, an environmental specification, as well as a set of measurable performance metrics."*

Any definition of a benchmark must be preceded by a detailed analysis of the application. This is traditionally not done since the architecture has been defined by hardware architects based on ad-hoc guessing. As a consequence many of these architectures have been grossly inadequate.

In the following we analyze the area of wireless communications and multimedia, which show similar properties. Commercially, these two areas are by far the most important ones for future MPSoC. The key metric for any mobile application is energy efficiency. The global performance measure is energy per decoded bit at the receiver. Clearly this measure is not practical since it describes the end result of the design process. This measure is a highly nonlinear function of the measures such as clock frequency, area, runtime, operations executed and power consumption. It may be mathematically viewed as a mapping of an N-dimensional space to a scalar quantity: energy efficiency.

To understand this mapping we need to analyze the application space. The requirements of mobile wireless communications and multimedia applications is qualitatively illustrated by Figure 7.1 which is due to Ravi Subramanian.

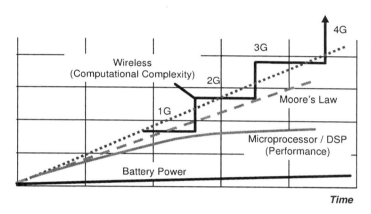

Figure 7.1. Algorithmic complexity growth in wireless applications (Source: R. Subramanian, Berkeley Design Automation Inc.).

The algorithmic complexity of advanced wireless communication systems increases exponentially in order to maximize the channel utilization measured in bits/sec per Hz bandwidth. Thus, the required computational performance measured in MOPS increases at least proportionally to the algorithmic complexity. This growth is far larger than what can be achieved by enhancements of general purpose DSP architectures. Until the year 2013, the International Technology Roadmap for Semiconductors (ITRS[1]) estimates an increase of performance requirements by a factor of almost 10,000. This cannot be addressed by the predicted clock frequency increase of a factor of five. Consequently there is a need for innovative computer architectures such as MPSoCs in order to achieve the required computational performance.

However, performance is only one of the drivers for MPSoC. The other is energy efficiency. It is well known that battery energy remains essentially flat.

---

[1] *http://public.itrs.net*

Thus, if the numbers of MOPS must increase exponentially, energy efficiency must at least grow with the same exponent.

This can be achieved only by heterogeneous multi-processor architectures. While the programming of complex multiprocessor architectures is an unsolved problem in general, taking into account a priori knowledge of the application promises to make the problem manageable.

In order to illustrate these aspects, the example of a UMTS receiver will be analyzed in the following (Figure 7.3). The signal processing task of a digital wireless receiver is characterized by the following properties.

- The individual tasks are loosely coupled. This eases the spatial mapping to application specific processing elements. These elements need to be matched to specific classes of sub-tasks, e.g. decoding.

- Usually, the task is periodic and only occasionally interrupted. This property allows the temporal mapping due to the pre-defined schedule.

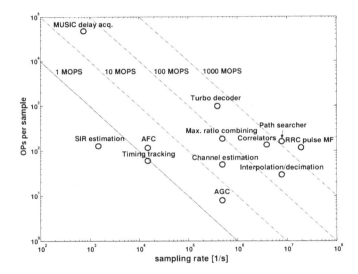

Figure 7.2. Computational complexity of a 384 kbps UMTS base band receiver.

On the horizontal axis we show the sampling rate of the periodically executed signal processing algorithm while on the vertical axis we show the number of operations per sample. The dashed diagonals are therefore the isoclinal lines of constant operations per second (MOPS). Several key observations can be drawn from Figure 7.2:

- The global MOPS are grossly inadequate to characterize the task. The same MOPS may be required for an algorithm demanding a high sampling

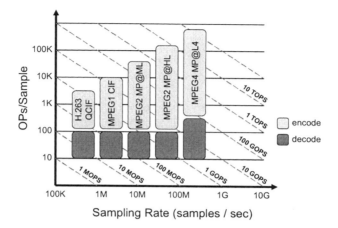

Figure 7.3. Computational complexity of video coding standards (Source: L.G. Chen, SIPS 2003).

rate but low complexity, or for a highly complex algorithm running at a low sampling rate. Examples for the former one are interpolation and decimation filters. The MUSIC delay acquisition is an example for the latter one.

- A large portion of the area in this figure is covered by different algorithms. It follows that a heterogeneous architecture is potentially an interesting candidate for this application. This architecture comprises a possibly large number of ASIPs each tailored to a specific class of algorithms.

Similar investigations have been performed on video encoding and decoding standards. In Figure 7.3 the performance requirements of the QCIF, MPEG1, MPEG2, and MPEG4 standards are depicted. It is clearly indicated that not only do performance requirements rise with improving standards, but that also the computational complexity ranges over several orders of magnitude. This analysis stresses again the vital need for heterogeneous MPSoC systems.

For simplicity reasons we have plotted a two-dimensional space. In reality, the space is three-dimensional with the additional axis being flexibility. In contrast to the first two dimensions flexibility is a very vague term, nevertheless it is of utmost importance. It is instructive to discuss the two extreme points of flexibility. On the one hand, maximum energy efficiency is achieved by a fixed functionality processing element, at the expense of zero post-silicon flexibility. On the other hand, maximum flexibility is achieved by a fully programmable device, at the expense of a very low energy efficiency. This is illustrated by Figure 7.4 which is due to Tobias Noll [26, 27].

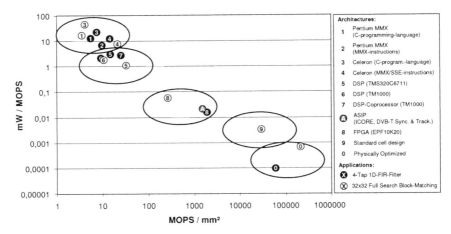

Figure 7.4. Comparison of algorithms on different architectures (Source: T. Noll, EECS RWTH Aachen University [26, 27]); Celeron®, Pentium®, Intel Corp.

In order to quantitatively compare architectures (apples to apples) in this figure, reciprocal energy efficiency (mW/MOPS) is plotted versus area efficiency (MOPS/mm^2). In Figure 7.4 energy efficiency is presented versus area efficiency for a number of representative signal processing tasks for various architectures. The conclusion that can be drawn from this figure is highly interesting. The difference in energy/area efficiency between fixed functionality and general purpose processors is orders of magnitude. An ASIP provides an optimum trade off between flexibility and energy efficiency not achievable by standard architectures/extensions.

So far we have discussed the spatial mapping of algorithms onto ASIPs. However, equally important is the temporal mapping since any MPSoC consists of processing elements and the communication among them. In the future we will find Networks on Chip (NoCs) rather than traditional bus concepts, since bus interconnects do not scale. Clearly, the temporal and spatial mapping must be supported with a programming model for the user to be useful in practice. This purpose is served by the NoC simulation framework GRACE++^{2} [129]. This approach enables the designer to become a virtual architect for the targeted MPSoC, as the complete system with the temporal and spatial mapping can be realized as a virtual prototype.

---

^{2}Developed at: Institute for Integrated Signal Processing Systems, RWTH Aachen University, *http://www.iss.rwth-aachen.de.*

2.   The Architectural Design Space

Element 2 – Inclusively Identifying the Architectural Space: *"A careful definition of the appropriate design space is essential for subsequent design space exploration."*

This element in the MESCAL approach applies especially for the selection of the appropriate architectures from the overall design space of MPSoC building blocks (Figure 7.4). In the design space, energy efficiency, performance, flexibility, and design efficiency have to be traded off between the extremes of highly flexible general purpose processors available off-the-shelf and fixed high performance full custom designs.

Today's applications have an increasing necessity for flexible solutions in order to improve the design efficiency and reduce the risk and costs of re-design cycles. Promising candidates for MPSoC building blocks that provide this flexibility are programmable application specific architectures. These so called Application Specific Instruction set Architectures (ASIPs) provide a high potential for flexibility, high performance, and dramatically improved energy efficiency through programmability and application specific optimizations.

Implementations of programmable architectures can vary widely and are closely linked to the application's requirements in terms of energy efficiency, flexibility, and performance. Therefore it is important to find the best starting point for the efficient ASIP design space exploration for a given application. Several architectural options are given in the next paragraph. Careful consideration of this decision for each individual building block of an SoC will result in a highly optimized heterogeneous MPSoC.

A traditional approach for a programmable architecture is the RISC concept, providing a single instruction issue pipeline. Superscalar architectures optimize performance by exploiting implicit parallelism within the sequential instruction flow. Very-Long-Instruction-Word (VLIW) architectures introduce explicit parallelism in the instruction word. Both superscalar and VLIW architectures trade off performance versus power consumption. A VLIW case study is given in Section 6.3. The issue of energy efficiency is targeted by the Single-Instruction-Multiple-Data (SIMD) approach as the data operation encoded in the instruction is executed on multiple data. This approach requires a high regularity of the data operations in order to exploit the SIMD potential. Therefore, applications handling more irregular data operations require the instructions to encapsulate fused operations [203]. These special instructions cover multiple application specific operations on multiple operands. This results in an architecture which provides high performance and high energy efficiency as proved in the case study given in Section 6.1.

The consideration of these heterogeneous architecture types is important for an optimum MPSoC implementation. In order to achieve a high design productivity at the same time, efficient ASIP design and exploration tools are inevitable. An efficient ASIP design flow must enable the designer to perform an efficient design space exploration by automatically generating consistent sets of simulators, software development tools, and optimized hardware implementations. Only with this approach will the cross-disciplinary task of ASIP design become broadly feasible. Thus, processor development will no longer be limited to few specialized companies. System houses will be able to do their own processor designs.

The Language for Instruction Set Architectures (LISA) is the language that the LISA Processor Design Platform is based on. This platform enables the designer to perform the desired efficient ASIP design as elaborated above. The LISA Processor Design Platform is described in detail in Section 4.

3.   ASIP Design Based on ADLs

ASIPs enable the designer to find the best compromise between flexibility, performance, and energy efficiency of a programmable solution. However, as these different goals have to be balanced, the ASIP design effort increases drastically. The following sections describe the traditional ASIP design methodology and motivate a highly efficient ASIP design approach based on Architecture Description Languages (ADLs).

3.1   Traditional ASIP Design Methodology

In the past, the long development time for ASIPs resulted from the fact that the design process was separated into four interrelated sequential phases.

- *Architecture Exploration:* The target application requiring application specific processor support is to be mapped efficiently onto a dedicated processor architecture. This process is composed of three main tasks. First, the application is profiled in order to determine critical portions that require dedicated hardware support. This task is referred to as hardware-software partitioning. Second, the instruction-set is defined on the basis of the hardware-software partitioning and profiling results. Third, an optimal microarchitecture that implements the instruction set has to be developed. These tasks of the architecture exploration cannot be performed in a top-down fashion. It is rather an iterative optimization process which is repeated until a sufficient fit between the selected architecture and the application is obtained. An architecture specific set of software development tools (assembler, linker, compiler, instruction set simulator, and profiler) is mandatory for that task. Each iteration in the exploration changes the architecture and leads to a time consuming and error prone change of these tools. The design of the compiler is an especially complex task. Therefore, its development was often omitted in past processor designs. This resulted in compiler-unfriendly architectures and the final application implementation was inefficient.

- *Architecture Implementation:* The processor specified in the exploration phase is implemented on RTL. Usually Hardware Description Languages (HDLs) like VHDL or Verilog® are used to develop a model which can be used as input for a standard synthesis flow. This manual implementation on RTL can easily lead to consistency problems between the textual architecture specification, the implementation, and the software development tools. Moreover, in contrast to the implementation with an HDL, the software tools are usually realized using a High Level Language (HLL) like C/C++.

- *Software Application Design:* Software application design requires a set of production-quality software development tools with different requirements than those used during the architecture exploration and implementation steps. For instance, during the implementation phase a cycle accurate simulator is mandatory. Because of the high degree of detailed architectural information this simulator is rather slow. In contrast to this, the focus for the Instruction Set Simulator (ISS) lies on execution speed rather than accuracy. The gap between these different requirements often leads to a complete re-implementation of the software development tool suite and therefore to consistency problems.

- *System Integration and Verification:* Co-simulation interfaces must be developed to integrate the software simulator for the chosen architecture into a system simulation environment. These interfaces vary with the architecture that is currently under development. Again, manual modifications of the interfaces are required with each change of the architecture, which is a tedious and lengthy task.

To further complicate this process, these four design phases are assigned to different design engineer groups with expertise knowledge in their respective fields. Design automation is mostly limited to the individual design phases, since even software tools and description languages vary from phase to phase. Thus, a very important but time-intensive factor is communication and verification between different design groups and design phases.

3.2   The ADL Approach

The development time can be decreased dramatically by employing an automated retargetable approach using ADLs. ADLs can be used to describe the architecture on a higher abstraction level than RTL. The levels of abstraction differ between the various ADLs. The two most important abstraction levels are the instruction-accurate and the cycle-accurate levels. While the instruction accurate model covers the behavior of the instructions without detailed timing information, a cycle accurate model also contains accurate information about timing and pipeline effects.

ADLs enable an efficient design space exploration by the automatic generation of the required software development tool kit. Very recent ADLs also enable the generation of synthesizable RTL descriptions. This approach makes it possible to use a single architecture description and therefore eliminates the consistency problems of the traditional ASIP design flow. Changes in the processor description directly lead to a completely new and consistent set of software tools and RTL implementation. This dramatically reduces the time which is required for each iteration cycle during the exploration. Thus, a larger number of architectural alternatives can be explored. ADLs can coarsely be grouped into two categories:

- *Instruction-set centric languages:* Instruction-set centric languages characterize the processor by its instruction-set architecture with the primary goal of retargeting HLL compilers. As the architecture information required for this purpose primarily concerns the instruction-set and constraints on the sequencing of instructions and instruction latencies, no further information on the micro-architecture needs to be included. Examples for this type of ADLs are nML [66], Sim-nML [196] and ISDL [92].

- *Architecture centric languages:* Architecture centric languages focus on the structure of the processor. Thus, they are practical for the mapping of architectures to RTL. However, extracting information concerning the instruction set from the structural model description in order to generate software tools is a complex task. MIMOLA [19] is an example for such an architecture centric language.

- *Mixed instruction-set centric and architecture centric languages:* Mixed ADLs describe both the instruction-set including the behavioral aspect of the processor and the underlying structure of the design. This enables the retargeting of the software tools as well as the generation of a hardware description from one ADL model. Mixed instruction-set centric and architecture centric languages are, e.g., FlexWare [184], EXPRESSION [93], and LISA [98].

4.   The LISA Processor Design Platform

The Language for Instruction Set Architectures (LISA) is the ADL which is used by the LISA Processor Design Platform for architecture specifications. LISA has been developed at the Institute for Integrated Signal Processing Systems at the RWTH Aachen University and is commercialized by CoWare Inc.[3] This platform targets *architecture exploration, architecture implementation, software tools generation* and *system integration*, which is congruent with the MESCAL approach.

Element 3 – Efficiently Describing and Evaluating the ASIPs: *"Comprehensive exploration of the design space requires software tools to map the application benchmark(s) onto the architectural design point in consideration, as well as to evaluate the quality of this mapping for the design metrics in consideration."* In Section 4.4 we explain the software tools generation and in Section 4.3 the hardware implementation from a single LISA model.

Element 4 – Comprehensively Exploring the Design Space: *"Design space exploration is a feedback driven process [...]. This is done using the appropriate software environment consisting of a compiler [...] and simulators to quantify the quality of the result. The key to doing this efficiently for a large number of design points is to quickly reconfigure the framework and simulators for each design point."* In Section 4.2 design space exploration based on the LISA language is presented.

Element 5 – Successfully Deploying the ASIP: *"An ASIP may achieve high efficiency by providing for an excellent match between the application and the architecture and at the same time may be completely useless if it cannot be programmed efficiently."* For this purpose, the compiler development is an important task in ASIP design. Its LISA based design is elaborated in Section 4.4.2. Also, automatically integrating LISA processor cores into the SoC environment for system simulation is possible on multiple levels of abstraction (Sections 4.5 and 7.1). However, mapping the complete application to the heterogeneous MPSoC platform down to a set of binary processor executables is a challenging task and still an area for active research. Several approaches are targeting this challenge, as for example our own work [125].

Only a single LISA architecture specification is necessary to perform all these tasks efficiently (Figure 7.5). The architecture description language LISA and each of these four tasks are described in the next five sections. As verification is necessary during architecture implementation, software application design, and integration, it is described separately in Section 5.

---

[3] *http://www.coware.com*

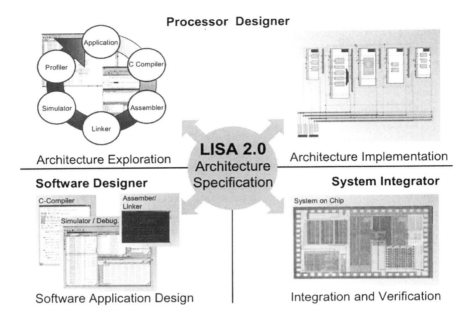

Figure 7.5. LISA development platform.

4.1 The LISA Language

This section briefly introduces the LISA 2.0 language. Using the LISA language, the target processor can be represented on several levels of abstraction, while refining the model in a successive fashion. The levels of abstraction encompass both the *structural* and the *temporal* accuracy, as will be discussed in the following.

During the architecture exploration phase, it is important to perform a seamless refinement of the processor description as the design space becomes narrower. LISA offers exactly this capability to the designer. The processor model can be described with LISA on different levels of abstraction, thereby allowing a seamless stepwise refinement. Abstraction can take place in two different domains: *architecture* abstraction and *temporal* abstraction.

As shown in Figure 7.6, the most abstract architectural level is the level of pseudo resources. Models on the finest abstraction level, here phase-accurate models, contain for example interrupts. Similarly, concerning temporal abstraction, an initial LISA description may start with pseudo instructions. They can be refined to processor instructions in an instruction-accurate description culminating to a phase-true instruction execution in the most accurate description of LISA.

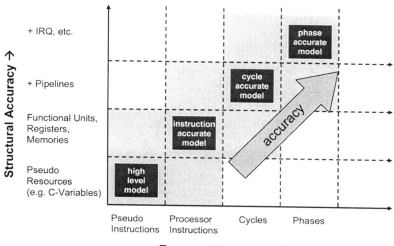

Figure 7.6. LISA abstraction levels.

The temporal refinement of the instructions leading to a more accurate model always requires a higher level of detail for the behavioral description. The *behavioral part* of a LISA description is able to cover the instruction-set regarding the instruction encoding, assembly syntax, functional behavior, and timing.

Besides this behavioral description of the architecture, its structure is of high importance. Especially for the generation of a hardware description this information is mandatory. Therefore a LISA model contains a *structural part* specifying the processor resources such as registers, memories, and pipelines.

4.2   Architecture Exploration

As indicated before, architecture design requires the designer to work in two domains (Figure 7.7): On one hand, the development of the software part including compiler, assembler, linker, and simulator and on the other hand the hardware implementation of the target architecture itself. The software simulator produces profiling data and thus helps answering crucial questions concerning the instruction set, the algorithm performance, and the required size of memories and registers. The required silicon area or power consumption can currently only be determined accurately in conjunction with a synthesizable HDL model.

The LISA exploration platform can generate the following tools:

- *C compiler:* The high level language C is commonly used for application programming. Therefore a C compiler is required in order to translate the C application to assembly code for the customized Instruction Set Architecture (ISA).

- *Assembler:* An assembler is required that translates human readable assembly instructions into binary object code for the respective ISA.

- *Linker:* The linker combines binary object code and data memory contents into an executable file with respect to the application specific memory architecture. It is controlled by a dedicated linker command file.

- *Instruction-set architecture simulator:* An Instruction-Set Simulator (ISS) provides extensive profiling capabilities like instruction execution statistics and resource utilization statistics already during the exploration cycle for both the architecture and the application.

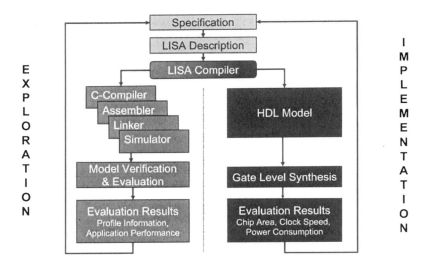

Figure 7.7. LISA architecture exploration flow.

Besides the ability to generate a set of software development tools, synthesizable HDL code (VHDL and Verilog®) can be generated automatically from the LISA processor description [210, 211]. This comprises the control path as well as the datapath. It is obvious that deriving both the software tools and the hardware implementation model from one single specification of the architecture in the LISA language has significant advantages: Only one model needs to be maintained and changes on the architecture are applied automatically to the software tools and the implementation model. Thus, the consistency problem among the individual software tools and between software tools and the implementation model is eliminated.

4.3   Automatic Architecture Implementation

Traditionally, the RTL implementation of an ASIP has been done by experienced designers. It is a complex and error-prone task that can affect the development cycle adversely. With the advent of ADLs, the automatic RTL generation from an ADL-based ASIP description became possible. Automatically generated RTL models proved to be useful for measuring the accurate processor performance during ADL-based architecture exploration. But the poor quality of the first generation RTL code compared to the hand-written RTL description prevented it from being used as a final implementation.

The hardware design platform based on LISA not only supports the automatic generation of an RTL description in several hardware description languages, but as a state-of-the-art tool, it also includes effective high-level optimization techniques [212] as well as the automatic generation of a hardware debug mechanism and a JTAG interface.

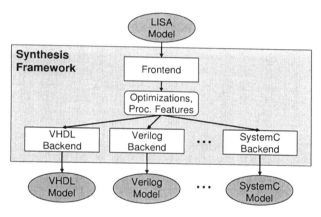

Figure 7.8.   Automatic RTL description generation from LISA.

These capabilities result from the architecture of the hardware synthesis framework depicted in Figure 7.8. The RTL generation from LISA is subdivided into three major phases. First, the LISA description is parsed by a frontend and translated into an intermediate architecture representation that keeps semantic information of the LISA description. On the basis of this representation, optimizations are applied and additional processor features are included depending on the desired configuration of the processor. The final phase is the generation of the RTL description itself. Multiple language-specific backends perform language-dependent code generation. This makes the complete generation process extremely flexible to adapt to new hardware description languages. Currently, VHDL, Verilog®, and SystemC™ are supported as RTL description formats.

The key motivation behind embedding high-level optimization techniques in ADL-driven automatic RTL generation is the availability of information about semantic relations and mutually exclusive execution which can hardly be extracted from an RTL representation. The optimizations possible with this information result in RTL descriptions whose gate level synthesis results prove to be of production quality and have been used in commercial SoC designs (cf. Sections 6.1 and 6.2). The most important optimization techniques are shortly described in the following.

- *Decision Minimization:* Decision minimization utilizes semantic information in order to move condition-independent code out of the surrounding conditions. This optimization reduces multiplexer instantiations and improves the timing significantly.

- *Signal Scope Localization:* In LISA, it is possible to declare and use a signal resource globally in the model. During signal scope localization, locality of the signal usage is explored and affected signals are converted into local resources.

- *Decoder Distribution:* Decoder distribution is a structural optimization where the instruction decoder is distributed over the entire pipeline. The decoded signals from earlier stages are fed into latter stages, in case they are used more than once.

- *Port Sharing:* During the port sharing optimization, the exclusiveness relations of resource accesses are considered by mapping mutually exclusive accesses to shared ports.

- *Resource Sharing:* Resource sharing is performed using the exclusiveness information and cost models for chip area and signal delays. Based on this information, and the cost models and constraints set by the designer, the sharing algorithm selects the sets of computational resources for sharing.

Besides optimization, the synthesis framework enables the automatic generation of processor features as all the required semantic information is available. One of these features is a hardware debug mechanism [209]. It enables the designer to debug software in its final hardware environment by giving access to the state of the processor core via an additional interface. The JTAG interface is commonly used for this purpose [105].

Implementing such a debug feature manually on RTL would be a lengthy and error-prone task, resulting in an implementation that provides no flexibility in case the requirements change. However, using the approach of automatically generating this processor feature, the designer is able to include necessary debugging capabilities into the target architecture in the exploration cycle as all the necessary flexibility is provided. Furthermore, the designer is able to explore the impact of the debugging mechanism on the processor characteristics like area, timing, etc. early in the design process. The debug mechanism is highly configurable and can therefore be adapted to the requirements of the specific design, avoiding the waste of physical resources.

4.4 Application Development Software Tools

The LISA processor design platform generates various high quality code generation and application development tools. Among these, the LISA simulator, based on the Just-In-Time Cache-Compiled simulation, and the C Compiler are described in this section.

4.4.1 The LISA Simulator.
The LISA simulator supports three different simulation techniques, which are described in the following.

- *Interpretive Simulation:* The interpretive simulation technique is a software implementation of the underlying decoder of the architecture. For this reason the interpretive simulation is considered to be a virtual machine performing the same operations as the hardware does: it fetches, decodes and, executes the instructions. All simulation steps are performed at runtime, which provides the highest possible flexibility. However, the straightforward mapping of the hardware behavior into a software simulator is the major disadvantage of the interpretive simulation technique. Compared to the decoding of the instructions in hardware, the control flow requires a significant amount of time in software.

- *Compiled Simulation:* The compiled simulation uses the locality of code in order to speed up the execution time of the simulation compared to the interpretive simulation technique. The task of fetching and decoding an instruction is performed once before the simulation runs. The decoding results are stored and reused later during simulation. Execution time is saved because during the simulation the fetch and decode steps do not need to be executed. Thus, the compiled simulation requires the program memory content to be fixed before simulation runtime. Various scenarios are not supported by the compiled simulation technique, such as system simulations with external and thus unknown memory content or operating systems with changing program memory content. Additionally, large applications, which require a huge amount of memory on the target host, are hard to support.

- *Just-In-Time Cache Compiled Simulation (JIT-CCS):* The objective of the JIT-CCS is to combine the advantages of both interpretive and compiled simulation. This technique provides the full flexibility of the interpretive simulation while reaching the performance of the compiled simulation. The underlying principle is to perform the compilation, in fact the decoding process, *just-in-time* at simulation runtime. Because of that, full flexibility is provided. The decoding results are stored in a software cache. In every subsequent simulation step, the cache is searched for existing decoding results. Due to the locality of code in typical applications, the simulation speed can be improved using the JIT-CCS. The cache size used in the JIT-CCS is variable and can be changed in a range from 1 - 32768 lines, where a line corresponds to one decoded instruction. The maximum amount of cache lines corresponds to a memory consumption of less than 16 MB on the simulator host. Compared to the traditional compiled simulation technique, where the complete application is translated before simulation time, this memory consumption is negligible.

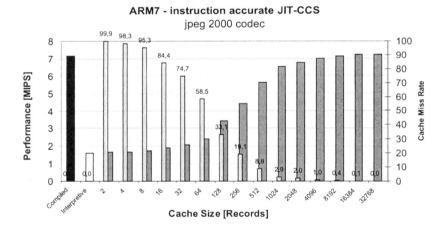

Figure 7.9. Performance of the Just-In-Time cache compiled simulation.

Figure 7.9 illustrates the performance of the cache compiled simulation of an ARM7[TM] core depending on the cache size. The results were achieved by using an 1200 MHz Athlon[TM] PC, 768 MB RAM running the Microsoft Windows[®] 2000 operating system. A cache size of one line means that the Just-In-Time cache compiled simulation essentially performs the same way as the interpretive simulation. Every instruction is decoded and simulated again, without using the advantage of code locality. With a rising number of cache lines the simulation performance (wide bars) comes closer to the performance of a compiled simulation and finally saturates. The simulation speed increases with the decreasing cache miss rate (narrow bars). As can be seen in the figure, the performance of the compiled simulation can be reached with a relatively small cache size and thus less memory consumption on the host machine. Moreover, this memory consumption on the target host is constant relative to the application size.

4.4.2 The Compiler Designer. Having a C compiler in the architecture exploration loop enables the software designer and architecture designer to study the application's performance requirements immediately after the algorithm designer has finished her/his work. Keeping the C compiler as part of the exploration loop also makes sure that the final architecture is compiler-friendly and development of a C compiler does not become an impractical task. Besides, a compiler/architecture combination can easily be adapted to further applications of the same application domain without the tedious and error-prone task of writing hundreds of lines of assembly manually.

Iterative architecture exploration demands a very flexible retargetable C compiler which must be quickly adapted to varying target processor configurations. LISATek™ relies on CoSy®[4] Express which is derived from the CoSy® compiler development system from ACE. CoSy®'s modular design, surrounding a generic, extensible intermediate representation (IR), offers numerous configuration possibilities both at the level of the IR and the backend for machine code generation. It is an industry proven compiler framework to develop production-quality performance compilers for a broad range of processor classes.

CoSy® generates compilers from so-called Code Generator Description files (CGD). A CGD model consists mainly of three components:

- Available *target processor resources* like registers or functional units.

- A description of *mapping rules*, specifying how C/C++ language constructs map to assembly instructions.

- A *scheduler table* describing instruction latencies and resource usage.

Apart from that, some further information like function calling conventions, C data type sizes, and alignment is required.

The LISATek™ Compiler Designer generates such a CGD description from a LISA processor model (Figure 7.10). This is a quite challenging task: Some information can be directly extracted from LISA models (e.g. processor re-

---

[4]CoSy is an international trademark of ACE Associated Computer Experts bv., *http://www.ace.nl.*

sources), but other information, e.g. the generation of an instruction scheduler, is only implicit and needs to be extracted by special algorithms. Other compiler-specific information like C-type bit widths is not at all present in the LISA model. Furthermore, the generation of C compilers has different requirements on instruction modeling than the other software tools. Compilers need a high level model of the target machine, whereas other tools like the simulator require detailed information about the cycle and bit-true behavior of machine operations.

Figure 7.10. C-Compiler Design Flow.

The LISATek Compiler Designer employs a semi-automatic approach to address these challenges instead of sacrificing flexibility or code quality. The Compiler Designer reads in a LISA model, extracts compiler relevant information whenever possible, and presents them in an easy-to-use GUI. The GUI is organized in different configuration components and the user is guided step-by-step through the specification of the missing items that could not be configured automatically.

- *Data layout, register allocator, and calling conventions:* Values for purely *numerical parameters* like C type bit widths, type alignments, and minimum addressable memory unit size not present in the LISA model can be entered in GUI tables.

  Several configuration items regarding the registers have to be defined. First of all, the user has to configure the *register allocator* by selecting the allocatable registers from the set of all available registers in the LISA model. Two dedicated registers for the stack and frame pointer also have to be defined. Registers which cannot be temporarily saved in the memory have to be excluded from the set of pushable registers.

  The *calling conventions* describe the state of the processor before calling a function and after returning from it. Configuration options include the specification of resources used to pass function parameters and return values. The GUI provides a convenient dialog to define for each C data type the preferred passing mechanism, which can be either register or stack.

- *Instruction schedulers:* Instruction schedulers decide about the sequence in which instructions are executed on the target processor. Beside structural hazards, data dependencies between instructions need to be taken into account. These constraints are captured by *scheduler tables* containing latency information for the different kinds of dependencies and the resource usage of instructions. These tables are generated automatically from the LISA model [237]. As the generator has to ensure a correct scheduler, it sometimes calculates conservative latency values. Therefore, the extracted scheduler characteristics can be manually overridden in the GUI in this case. From this information an improved list scheduler is generated, capable of efficiently filling delay slots.

- *Code selector:* In order to get an operational compiler a minimal set of code selector rules, i.e. the mapping of compiler IR operators to assembly instructions, is needed. Like in most compilers, these mapping rules are the basis for the tree pattern matching based code selector in CoSy®.

To quickly get a first fully working prototype of the compiler these mappings can be automatically generated from a LISA model enriched with information about the semantics of instructions [41, 33]. The code quality produced by such a compiler is acceptable considering that it is actually an out-of-the-box compiler without any target specific optimizations. However, to achieve better code quality, the user may refine the generated code selector rules listed in the GUI manually, or he can add more dedicated mapping rules that efficiently cover special cases which have not been considered by the mapping generator. This can be done within the so-called *mapping dialog* (Figure 7.11) [100].

The mapping dialog presents in the top left part the set of available IR operations as well as the hierarchically organized set of machine operations in the given LISA model (right window).

By means of a convenient drag-and-drop mechanism, the user can compose mapping rules (top center window) from the list of IR operators. For a particular pattern an appropriate instruction has to be selected from the list of machine operations. Finally, the arguments of the mapping rule have to be linked via drag-and-drop to the operands of the machine operations. This way, complex mapping rules covering, e.g. a MAC operation, can be easily specified. It is also possible to assign several instructions to rules which could even contain control flow.

Based on the selected (either manually or generated) machine operations for a mapping rule, the compiler generator looks up the corresponding assembly syntax (bottom center window) in the LISA model and can therefore automatically generate the code emitter for the respective mapping rule.

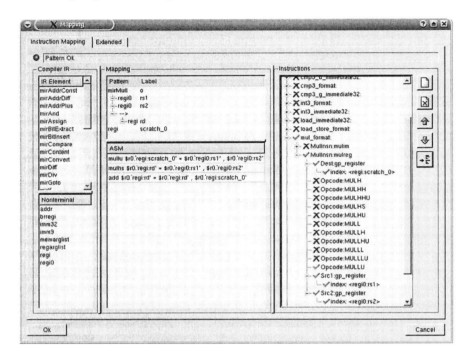

Figure 7.11. GUI mapping dialog.

The final output of the Compiler Designer is a CGD compiler specification file, which finally is further processed by CoSy® to generate the C-compiler.

4.5 System Integration

Today, typical single chip electronic system implementations combine a mixture of DSPs, micro-controllers, ASICs, and memories. In the future, the number of programmable units in a System-on-Chip (SoC) design will drastically increase making "processors to be the transistors of future SoCs." To handle the enormous complexity, system level simulation is absolutely necessary for both performance evaluation as well as verification in the system context. The earlier design errors or performance shortcomings are detected in the design flow, the lower the cost for redesign cycles becomes. The automatically generated LISA processor simulators can be integrated into various System Simulation

Environments, such as CoWare ConvergenSC™ [5] or Synopsys CoCentric System Studio[6]. Thus, modules provided by different design teams or even third parties can be combined easily.

The communication of the LISA processors with their system environment can be modeled on different levels of abstraction. For pin accurate co-simulation, LISA pin resources can be mapped directly to the SoC environment. Alternatively, the LISA bus interface allows modeling the SoC communication on a higher abstraction level, i.e. Transaction Level Modeling (TLM) [89]. By that, accesses to buses and memories external to the respective processor core are efficiently mapped onto the communication primitives applied in the SoC simulation environment. Automatically generated wrappers allow integration of fast LISA processor simulators into the final system context for the abstraction levels introduced in Section 7.2 [242].

For user-friendly debugging and online profiling of the embedded SW and its platform, the user always has the possibility of getting the full SW centric view of an arbitrary SW block [243] at simulation runtime. This is done just by dynamically connecting a multiprocessor debugger GUI to the processor to focus on. All other SW blocks that do not need to be inspected still simulate at maximum speed without a GUI connection. The remote debugger frontend instance offers all observability and controllability features for multiprocessor simulation as known from stand-alone processor simulation. Even resources external to a processor module but mapped into its address space like peripheral registers and external memories can be visualized and modified by the multiprocessor debugger GUI (Figure 7.12). The SW developer can dynamically connect to particular processors, set break/watch points in respective code segments, disconnect from simulation, and automatically re-connect when a breakpoint is hit.

5. ASIP Verification

Verification is an essential part of any processor design. No matter what design flow is actually chosen, implementations have to be verified against specifications in many phases of the design. Although the LISATek approach reduces the verification effort by automating large parts of the design flow from a high-level functional processor model down to the lower level of a synthesizable hardware implementation, there is still need for comprehensive verification. Moving from one abstraction level down to another one, e.g. from instruction accuracy down to cycle accuracy, the designer is required to add additional implementation details, i.e. potential sources of errors. Even if the

---

[5] *http://www.coware.com*
[6] *http://www.synopsys.com*

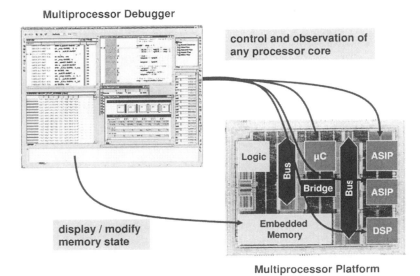

Figure 7.12. Principle of Multiprocessor Debugging during System Integration.

generated hardware implementation is assumed to be correct by construction, there might be need for manual modifications and optimizations which again require verification.

Having a high level description of the ASIP that can serve as an executable specification enables a simulation-based verification approach. This means that the simulation result of the high level processor model is taken as a golden reference and compared with results produced by the model under test. Changes in the processor state need to be traced and compared with each other, e.g. register contents, read/write accesses to memory, status flags, and pipeline states. The LISATek™ tool suite provides extensive tracing capabilities, including the creation of Value Change Diagrams (VCDs) similar to those created by RTL simulation tools. The VCDs of the reference model and the model under test can then be directly compared. The model is considered to have an equivalent behavior if the changes in state due to the stimuli are equal in both models, as depicted in Figure 7.13.

Creating tests is a challenging and time consuming task that requires a lot of expert knowledge. Designers usually create tests by hand for the new features they implement. Usually, a lot of expert knowledge is inherent to those tests that often include corner cases that are hard to reach. Hence they should be preserved and collected in order to enhance the regression sets. Tests that have discovered bugs should also be kept in the regression set preventing bugs from returning that have once been fixed. In some cases, there are commercial test

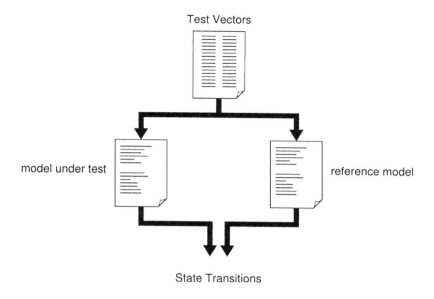

Figure 7.13. Simulative verification.

suites available that test special functionality of the design. For instance, if the ASIP is used to realize the protocol stack in a networking environment, there are most likely test cases available from the standardization committee to check that functionality including every corner case. Due to the uniqueness of the design, there will always be bugs that cannot be found with a predefined test suite. In every case, massive random testing with high coverage is necessary. In order to achieve a high coverage, huge numbers of tests are required, resulting in a substantial simulation time. In cases of designs with considerable complexity, testing can even become a major bottleneck in the design process. Thus it is essential to intelligently create the random test vectors such that redundancy is avoided. Tests created randomly heavily stress parts of the design while other parts are not tested at all. Tests generated intelligently balance the load between the parts of the design. The achieved level of confidence depends on the quality of the test vectors. The time needed for verification depends on the number of test vectors. Thus, the most crucial part of verification is to find test cases with a high coverage of the processor's functionality but with low redundancy.

5.1 Automatic Generation of Test Vectors from LISA Descriptions for Instruction Set Verification

High level processor descriptions that abstract from implementation details make it possible for analysis tools to automatically identify key functionalities of the design, enabling automatic generation of well balanced test cases. In a LISA description, resources like register sets, pipelines, and memories are explicitly instantiated and the instruction set is directly described. This kind of information would be impossible to extract from an RTL implementation of a processor. A simple, yet efficient way of test pattern generation is to enumerate all legal instructions. Immediate fields of the generated instructions that carry data but do not contribute to the instruction type can be randomly filled. This way, an arbitrary number of random tests can be generated that are well balanced over the entire instruction set.

A more advanced way of automatic test generation has been described in the work of Lüthje [155]. Here, not only the instruction set itself, but also the behavior of the instructions is taken into account. The embedded C code in a LISA description that describes the behavior of the instructions is analyzed with the newly developed analysis technique named *abstract execution*. The basic concept of abstract execution is to interpret the code to be analyzed without having full information about the data being processed. Instead of concrete values being processed, collections of information about the values are used. An example of this kind of information is in the case of a numeric operand the *range* of possible values. The lack of full information about the operands during interpretation can cause ambiguities in the control flow. In these cases, the different branches of the control flow ambiguities are executed in parallel. Changes to contents of memory in the different branches yield alternative contents in memory that are stored together with the conditions under which they are taken.

In order to analyze a processor model, it is interpreted while processing one single instruction. Doing this, neither the instruction itself nor the processor state, i.e. the contents of the resources of the processor model, are known. Therefore, the execution of this one "virtual" instruction represents any possible behavior of the processor model. During interpretation, the desired information about the processor model is collected. It is further described how these results can be used to design a set of equations whose solutions directly lead to a set of test cases that guarantee a full code coverage of the model. For a medium complexity model, this is usually not more than a few hundred test cases. However, the number of tests can be multiplied arbitrarily by finding more than one solution for each of those highly underdetermined equations. This way, perfectly balanced tests can be generated that spread over the entire functionality of the model.

5.2   Genesys

Genesys was originally devised to cope with very complex large-scale processor systems [3, 70]. It has a variety of capabilities targeted at verifying complex mechanisms, such as Memory Management Units (MMU - virtual memory), cache protocols and hierarchies, multi-processor configurations, and so on [205]. Since its development it has successfully been applied to IBM products all over the world. Beyond that it has received recognition by non-IBM customers as well. In particular, Genesys has successfully been used for the verification of IBM's PowerPC® and Intel's X86 family. Although Genesys is originally targeted at large-scale processor families, IBM claims it to be also beneficial for verification of DSPs and ASIPs.

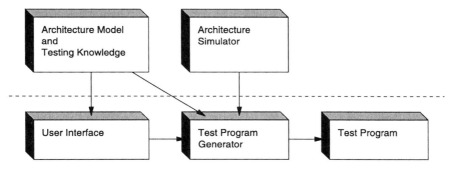

Figure 7.14.   Genesys building blocks.

Figure 7.14 shows the building blocks of the Genesys system. The architecture model, the testing knowledge, and the simulator are architecture-specific and have to be provided by the user. The user describes the processor's instruction set, resources and, data types. The simulator is an instruction accurate simulator that serves as a reference model. Besides the typical simulator features, it implements a special Genesys API. Furthermore, it can automatically be created from a LISA processor model by the LISATek™ tool suite. The Genesys system provides the following special functionalities:

- Random access to any processor resource, in particular memories and registers.

- Tracing reports about read/write accesses to processor resources.

- Rollback functionality to restore arbitrary processor states during simulation.

The instruction set is represented by instruction trees. Each instruction tree specifies the general structure of the corresponding instruction. During test generation, the instruction trees are used to create generic instructions. The rules stored in the knowledge base (testing knowledge) are then used to specialize these instructions such that specific properties or corner cases of the architecture are tested. The rules in the knowledge base contain expert knowledge and have to be set up by testing experts. For instance in connection with an ADD operation, a rule could cover the corner case of the result being zero. It would have to make the operands equal but with opposite signs. The architecture simulator is needed to predict the results of the instruction execution.

Test programs are generated by the test generator applying two principles. Instructions are generated as described above on the basis of the architecture model and the testing knowledge. However, if no complete test specification is available, random decisions are made in order to complete test scenarios. This way, an unbounded number of different tests can be generated, but unlike in fully random test generation, tests are balanced. Thanks to the testing knowledge, the probabilities of interesting corner cases can be controlled by the user. In particular, test generation can be influenced in the following ways [205]:

- The set of instructions can be selected that is to be used in the generated test program.

- The probability of exceptions can be specified.

- Sequences of instructions can be generated that repeatedly (re)-use the same resources.

- The user may specify functions that generate the operands in test instructions.

- Test patterns can manually be specified, e.g., to drive the system into a special state.

- There is special support for defining test scenarios for all kinds of loops.

- There is special support for exhaustive testing of parts of the architecture. For example, all possible data of an operand or all possible exceptions may be enumerated.

Thus, Genesys is a pseudo random test generator that allows the user to bring in his testing knowledge in order to improve test quality and reduce the number of required tests. However, the quality and coverage of the generated tests depends on the user's input. Genesys requires the architecture model to be transformed into a different representation at a high level of abstraction. It is well suited for complex multi purpose processors that it was originally developed for. Moreover it can also be applied to special purpose architectures.

6. ASIP Design Case Studies

In the following, several case studies with varying emphases are presented proving the feasibility of the LISATek™ approach. The example of the ICORE architecture focuses on deriving optimized application-specific instructions. The ASMD and LISA 68HC11 architectures demonstrate the modeling efficiency by improving existing architectures. Finally, the ST200 case study covers the modeling of a high-end VLIW architecture.

6.1 ICORE: Deriving Optimized Instructions

The ICORE [79] architecture has been developed for synchronization and channel estimation tasks in flexible and energy efficient digital receiver chips for terrestrial digital video broadcasting (DVB-T). It is initially used in a commercial set-top box. In order to reuse the ICORE architecture in handhelds running digital video broadcasting applications (DVB-H), energy optimizations have been of utmost importance. This processor was initially based on a mainly conventional DSP instruction set of a typical load/store Harvard architecture. The instruction set includes instructions for arithmetic and logical operations, data moves, and program flow control. This basic architecture has been implemented using the LISA design environment in order to tailor the architecture to the DVB-T application. Obviously, the first ASIP specification in general as well as the first ICORE realization are not yet optimal solutions with respect to execution time and energy consumption and might violate given constraints.

When profiling the first version of the architecture and its application, one important measure is the execution count for assembly instructions of the algorithm implementation. The execution count for every assembly address is given in Figure 7.15. The profiling clearly indicates the *CORDIC* subroutine as the hot spot of the application.

Energy efficiency was one of the primary goals in the ICORE project. In principle, the following energy optimization steps are applicable in order to maximize energy efficiency:

- High instruction coding density/application specific instruction encoding.

- Datapath optimization using application specific instructions and functional units.

- Local/global clock gating.

- Blocking logic to suppress wasteful logic and wire activity.

- Sleep/doze modes.

- Software optimization.

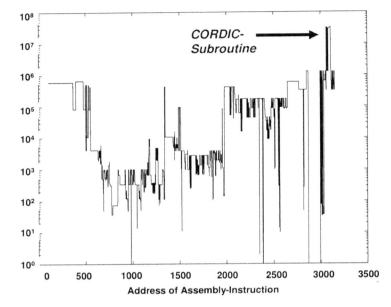

Figure 7.15. ICORE application profiling.

During the development of the ICORE architecture, it turned out that application specific instruction set optimizations are most effective for the overall reduction in energy consumption. According to Figure 7.16, the overall energy consumption can be reduced by concurrently executing arithmetic operations [80]. The energy consumed by the useful arithmetic operations is referred to as *intrinsic energy*. This kind of energy consumption is independent of the operation schedule. The energy required to control the arithmetic operations, such as to load instructions, to decode instructions and to steer the control path, is called *overhead energy*. This kind of energy is strongly dependent on the instruction set and schedule.

The concept of concurrent execution of heterogeneous data-independent arithmetic operations evolved from an already known technique to increase throughput, the SIMD (Single Instruction Multiple Data) architectures. Such architectures perform the same arithmetic operations on multiple data values at a time. This concept is especially applicable to matrices and vector calculations. The instruction set optimization targeted in ASIP design eliminates the necessity to perform the same arithmetic operations on multiple data, but executes arbitrary arithmetic operations on multiple data, called *instruction*

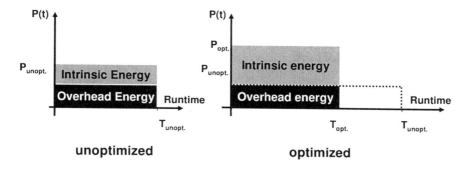

Figure 7.16. Energy optimization principle.

fusion (Section 2). In the following this approach will be elaborated for a manual analysis. However, recent research is applied in order to establish semi-automated and automated instruction set synthesis [10, 53, 122] (see also Tensilica's XPRES™ compiler in Chapter 8).

As indicated, the *CORDIC* subroutine has been identified as a hot-spot in the ICORE application. The arithmetic instructions which perform the *CORDIC* angle calculation have been reduced from five to only two instructions. Therefore, this optimization strongly reduces the energy consumption due to the reduced instruction fetch and decoding. Notice that the arithmetic calculation remains unchanged; only the scheduling of arithmetic operations is changed.

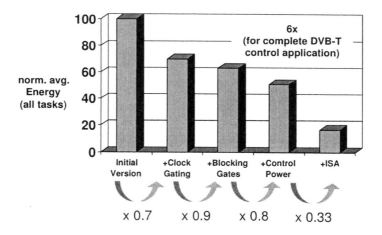

Figure 7.17. Overall energy reduction.

The overall energy consumption was reduced by a factor of six, as shown in Figure 7.17. This figure also clearly demonstrates that a much higher reduction is achieved by the instruction set optimization (a factor of 0.33) compared to traditional VLSI optimization techniques. In the ICORE project, instruction set optimization has been performed manually, while current research activities focus on an automatic C code analysis and instruction set generation.

In the following, the process of specifying and including one of the specialized *CORDIC* instructions in the ICORE LISA model is described in detail. The left hand side of Figure 7.18 shows the main loop of the *CORDIC* application (_COR_LOOPSTART to _COR_LOOPEND). It contains five instructions: one move instruction, three instructions performing arithmetic operations, and one instruction used for memory access. These instructions do not show direct data dependencies and therefore are ideal candidates for parallel execution.

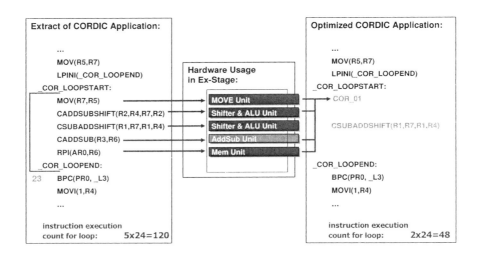

Figure 7.18. Instruction Set Architecture optimization for the CORDIC application.

Since three instructions utilize the ALU, a parallel execution is impossible without architecture changes. An additional calculation unit is required. CADDSUBSHIFT and CSUBADDSHIFT utilize both the ALU and the shifter unit. CADDSUB utilizes only the ALU performing either an addition or a subtraction depending on the required sign. Implementing an additional dedicated AddSub unit enables the parallel execution of an arbitrary ALU operation and an addition/subtraction in parallel, while keeping the additional hardware effort as low as possible. Figure 7.19 shows the implementation of this AddSub unit in LISA.

```
OPERATION AddOrSub IN pipe.EX
{
  ...
  BEHAVIOR {
    if((IN.Sign & 0x80000000)!=0) {
      // Perform the subtraction
      OUT.ResultC = IN.OperandE - IN.OperandC;
    }
    else {
      // Perfrom the addition
      OUT.ResultC = IN.OperandE + IN.OperandC;
    }
  }
}

RESOURCE
{
  ...
  UNIT Addsub { AddOrSub; };
}
```

Behavior description
of the new unit

Declare new unit

Figure 7.19. Implementation of an AddSub unit in LISA.

Depending on the read from the previous pipeline register element *Sign*, the result of an subtraction or addition of the operands *OperandA* and *OperandB* taken from the previous pipeline register is written to the element *ResultC* from the next pipeline register. In addition to the behavior of the unit, the unit itself has to be declared, which is done using the keyword UNIT as shown in Figure 7.19. Now, with the introduced AddSub unit, it is possible to map the instructions of the main loop on units in the execute stage as is depicted in Figure 7.18. Only the shifter and the ALU unit are used twice in one loop iteration. This enables the reassembling of four of the five instructions to a new specialized instruction, the COR_01 instruction, utilizing the units MOVE, Shifter, ALU, AddSub, and Mem.

Figure 7.20 shows the implementation of this instruction in LISA. First, the new coding and syntax of the new instruction are defined. Then, the behavior of the instruction is defined. Required operands are loaded from the register file and passed to the correct pipeline register element. The destination registers are specified by writing a corresponding number to pipeline register elements dedicated for this purpose. Also the new address for the memory access is calculated. Finally, the required and already existing datapath has to be utilized and triggered for execution. This is done by simply activating the corresponding LISA operations as shown in Figure 7.20.

```
OPERATION Cordic01 IN pipe.ID
{
  ...
  CODING{ 0b0111000 0bx[14] }                      ⎤   Define
  SYNTAX{ "COR_01" }                               ⎬   Coding & Syntax
  BEHAVIOR {                                        ⎦   of new Instruction
    OUT.OperandA      = R[2];                 ⎞
    OUT.RegisterNumA  = 7;                    ⎟
    OUT.OperandB      = R[4];                 ⎟
    OUT.RegisterNumB  = 2;                    ⎟
    OUT.OperandC      = R[6];                 ⎟
    OUT.RegisterNumC  = 3;                    ⎬   Load operands
    OUT.OperandD      = R[7];                 ⎭   and calculate new
    OUT.RegisterNumD  = 6;                        address
    OUT.Sign          = R[1];                 ⎟
    ...                                       ⎟
    OUT.Address = AR[0];                      ⎟
    AR[0] = AR[0] + 1;                        ⎠
  }
  ACTIVATION { Mem, SR_AddSub, AddOrSub, MOV }     ⎤   Utilize and trigger
}                                                  ⎦   existing data-path
```

Figure 7.20. Implementation of an additional, specialized CORDIC instruction in LISA.

Using the new COR_01 instruction as depicted on the right hand side of Figure 7.18 results in a reduction of required cycles for the total execution of the loop from 120 down to 48. This not only means a speed up in the execution of the application, but also an energy reduction as described before (Figure 7.16). This example shows that the implementation of additional units and instructions with LISA is straightforward and achieved with very little design effort.

The final ICORE is a typical 32-bit load-store Harvard processor architecture with an instruction issue rate of one. The structure of the core is depicted in Figure 7.21. It currently implements about 60 DSP instructions. There are 18 arithmetic instructions and 14 instructions for program flow control including zero overhead loop instructions. The remaining instructions are used for memory and interface I/O operations, bit manipulations, and logical operations. Highly optimized instructions support a fast *CORDIC* angle calculation, which have been derived from the optimization process described above.

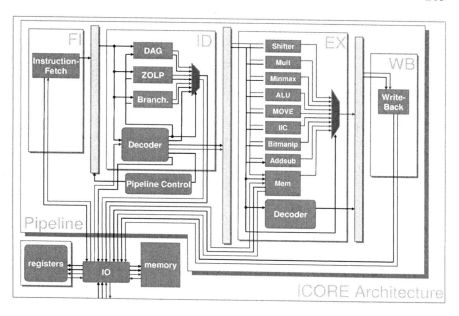

Figure 7.21. Structure of the ICORE architecture.

6.2 ASMD & LISA 68HC11: Improving Existing Architectures

The Application-Specific Multirate DSP (ASMD) from Infineon Technologies[36, 227] is a simple and small ASIP dedicated to interpolation and decimation. The structure of the ASMD is depicted in Figure 7.22.

The datapath of the ASMD consists of two shifter units, a cut unit, two inverter units, an adder, and two saturation units. Operands can be taken from state registers, arithmetic registers, and a constant array. The constants are hardwired for each implementation of the ASMD and therefore synthesized as combinatorial logic. Because of the use of these constants as fixed coefficients, it is possible to avoid the implementation of a full multiplier by using shifters and an adder. Due to the two implemented shifters each operand can be shifted separately. The cut unit enables the zeroing of an arbitrary number of least significant bits. Two inverter units in conjunction with the adder can be used to perform additions and subtractions. One saturation unit is used to reduce the word length of the input for the state registers and the other one for the output of the ASMD.

Filter applications for the ASMD are written in assembly. Since the filter does not have to be changed in its final environment, the application can be fully

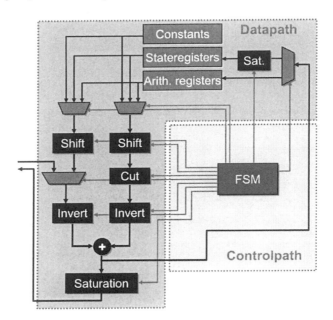

Figure 7.22. Structure of the ASMD architecture.

integrated into the final design. Therefore, the program memory content is not stored in a RAM or ROM but synthesized as combinatorial logic. This leads to an implementation of the control path as an FSM as depicted in Figure 7.22. The ability to use assembly programs for the implementation makes the development and adjustment of filters easy, fast, and intuitive.

The ASMD LISA model is derived from an existing RTL version of the ASMD developed in VHDL. The generated RTL description is synthesized with the Synopsys Design Compiler®[7] and mapped to a technology library, which is based on a 0.18 μm process. The synthesis results of the ASMD are presented in Table 7.1. The completely generated SystemC™ model fulfilled all design constraints and thus was taken for the final implementation.

The LISA model benefits from a very short design effort and design time for changes and extensions to the architecture. This enables an easy reuse of the architecture, as shown in the following. The evaluation of the ASMD with respect to a Bluetooth™ application uncovered the requirement of one additional instruction to increase the throughput significantly. For that reason the LISA model was changed, the RTL model regenerated and the new models verified within one day. Due to the large reduction in design time and the negligible

---

[7] *http://www.synopsys.com*

Table 7.1. Performance of the automatically generated ASMD implementation.

	Original ASMD	Generated ASMD	Ratio
Timing	4.22 ns	5.09 ns	1.21
Gates	9549 Gates	11678 Gates	1.22

overhead in speed and area, the generated version of the ASMD replaced the existing core and is now used in an Infineon Bluetooth™ device.

The 68HC11 architecture[8] is a well known 8-bit microcontroller from Motorola. The goal was to reuse legacy application code for a Bluetooth™ application while increasing the performance of the architecture. Thus, an architecture was developed that is compatible with the original architecture on the *assembly level*. The existent application and compiler were reused, whereas the assembler, linker, and the new HDL description were generated automatically from the LISA model.

While developing the LISA 68HC11 architecture, state-of-the-art architectural features and modern design aspects have been incorporated. The architecture is completely different from the original M68HC11 architecture. Unlike the original architecture, the new one is pipelined. It has three pipeline stages, namely *fetch*, *decode*, and *execute*. The pipeline has bypasses and is fully interlocked.

The original architecture has 8, 16, 24, 32, and 40 bit instructions. The number of cycles required to fetch the instructions depends on the width and therefore ranges from 1 to 5 cycles. Operand fetching may take additional cycles beyond the execution of the instruction itself.

In order to speed up the architecture, instruction and data memory were separated (Harvard architecture) and the bus width was increased from 8-bit to 16-bit (Figure 7.23). The coding of the architecture was reorganized to achieve a higher instruction throughput compared to the original architecture. The instruction set contains 16 and 32-bit instructions and is compatible to the original instruction set on assembly level while keeping the number of 32-bit instructions as low as possible. Instructions take one and two cycles respectively to be fetched. The fetch unit is responsible for reorganizing the 16-bit bundles to 32-bit instructions decoded in the second pipeline stage. Because of the Harvard architecture no additional cycles are required in order to fetch operands. The pipeline hides additional cycles for the execution as long as no interlocking is required. Exceptions are multiplication, division, and interrupt handling

---

[8]Motorola 68HC11 Microcontroller, *http://www.motorola.com*.

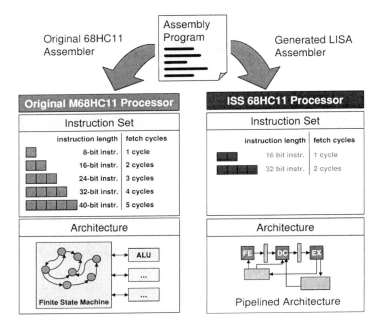

Figure 7.23. 68HC11 architecture comparison.

instructions, which require a multi-cycle execution. The speedup achieved by this implementation at the same clock frequency is about 60%.

Of key importance is the design efficiency for this architecture. The overall design time for the LISA 68HC11 processor was very short. The architecture has been developed and optimized within five weeks from instruction set definition, optimization, and architecture implementation to verification. The architecture was generated completely in VHDL without performing any manual changes to the generated code. The automatic optimization techniques introduced in Section 4.3 were used. Figure 7.24 shows the AT-chart containing several different reachable points in the design space using different optimization configurations.

A synthesis with Synopsys Design Compiler® using a 0.18 μm library resulted in a timing of 6.74 ns and a gate count of 16.64k gates. The synthesis results achieved with the automatically generated RTL model of the LISA 68HC11 are comparable with those of the handwritten implementation referring to the 68HC11 DesignWare® component. The latter occupies between 15 and 30k gates depending on the speed, configuration and target technology[9]. It

---

[9]Synopsys DesignWare® Components, *http://www.synopsys.com.*

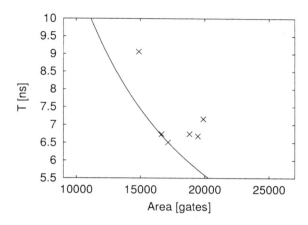

Figure 7.24.   AT-Chart for the LISA 68HC11 architecture.

is designed for applications running at clock frequencies up to 200MHz using a 0.13 μm process.

6.3   ST200: Modeling a High-end VLIW Architecture

This example illustrates the capability of LISA to model advanced computer architectures and outlines the achieved modeling efficiency.

ST200 is a family of high-performance low-power VLIW processor cores designed by STMicroelectronics for the use in SoC solutions running computationally intensive applications as a host processor or as an audio or video processor. Such applications include embedded systems in consumer, digital TV, and telecommunication markets.

6.3.1   ST200 Family of VLIW Cores.
The ST200 is a four issue VLIW processor which is significantly simpler and smaller than an equivalent four issue superscalar processor. The compiler is key to generating, scheduling, and bundling instructions, removing the need to add complex hardware to achieve the same results.

The ST200 design originates from the Lx research project started as a joint development by Hewlett-Packard laboratories and STMicroelectronics [65].

Figure 7.25 displays the basic organization of an ST200 processor. This processor executes up to 4 instructions per cycle, with a maximum of one control instruction (goto, jump, call, return), one memory instruction (load, store, prefetch), and two multiply instructions per cycle. All arithmetic instructions operate on integer values with operands belonging either to the general register file (64 × 32-bit) or to the Branch Register file (8 × 1-bit). The multiply

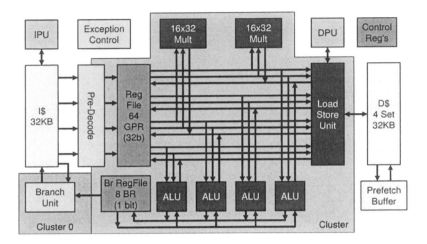

Figure 7.25. The ST200 VLIW processor architecture [65].

instructions are restricted to 16 by 32-bit on the earlier core variants, but have been recently extended to 32-bit by 32-bit integer and fractional multiplication on the ST231 core. There is no divide instruction but there is a division step instruction suitable for 32-bit division. In order to reduce the use of conditional branches, the ST200 processor also provides conditional execution of instructions.

The ST200 features a clean, compiler-friendly six-stage instruction pipeline, where all instructions are fully pipelined. Almost all instruction latencies are one cycle, meaning that the result of an instruction can be used in the next bundle. Some instructions have a three-cycle latency (load, multiply, compare to branch) or in one case a four-cycle latency (load link register LR to call).

6.3.2 VLIW Bundle Composition. As mentioned earlier, the ST200 design requires the availability of application code generation tools that are able to fully exploit the parallelism of the architecture. As the ST200 does not feature a dynamic instruction scheduling as found in superscalar architectures, it is the compiler's task to create an optimum instruction schedule by composing so-called bundles (VLIWs) from 32-bit instructions (further on called *syllables*). Constraints on how a bundle can be composed have to be considered during this process.

The ST200 uses a differential encoding of the bundles, i.e. each syllable encoding has a bit that indicates whether the bundle is complete or is followed by another syllable. This obsoletes the need to explicitly insert NOP instructions in case a bundle cannot be filled with sensible instructions, which gives a significant reduction in code size over classical VLIW architectures.

6.3.3 VLIW Modeling with LISA.

The LISA language allows a concise, modular, and readable description of the ST200 processor, as it allows to separate the description of bundle composition from the description of the syllables. An example for the description of a VLIW instruction word in LISA is given in Figure 7.26. This example demonstrates that not only the composition and full specification of VLIW bundles is possible within few lines (bdl_4), but that it is also possible to implement bundles with different numbers of syllables simply by extending the coding root (bdl_1).

```
DECLARE {
  ENUM   type = { bdl_1, bdl_4 };
  GROUP  syl1 = { control_op || alu_op || ldst_op || mul_op };
  GROUP  syl2 = { alu_op };
  GROUP  syl3 = { alu_op || mul_op };
  GROUP  syl4 = { alu_op };
}

SWITCH (type) {
  // Bundle only contains a single syllable
  CASE bdl_1: {
    CODING { 0b01 syl1 }
    SYNTAX { syl1 ";" }
    ACTIVATION { syl1 }
  }
  // Full bundle with the maximum of 4 syllables
  CASE bdl_4: {
    CODING { 0b00 syl1 || 0b00 syl2 || 0b00 syl3 || 0b01 syl4 }
    SYNTAX { syl1 || syl2 || syl3 || syl4 ";" }
    ACTIVATION { syl1, syl2, syl3, syl4 }
  }
}
```

Figure 7.26.   LISA description of the VLIW composition.

In addition to instructions, syllables can also be made up by long immediate values. In the ST200 architecture, all instructions supporting this immediate format can use 20-bit immediate values in either the previous or the next syllable relative to their own position in the bundle. Together with a 12-bit immediate value that is carried within the instruction itself, this format supports instructions with 32-bit immediate values. Naturally, a slot that contains such an immediate extension cannot carry an instruction.

Similarly to the four-syllable VLIW bundle composition in Figure 7.26, an example for instructions using an extended immediate field is given in Figure 7.27. Here again, the coding root is simply extended by specifying the VLIW bundle bdli. In this bundle the coding of the LISA operation long_op

spans two syllables. This LISA operation encapsulates all instructions that require such long immediates. Therefore all LISA operations below the coding
root, including long_op, are independent from the actual VLIW composition
and VLIW implementation. Thus, the VLIW composition can always be extended without the need to modify any other part of the LISA model.

```
...

CASE bdli: { // Bundle with long immediate syllable
  CODING { 0b00 syl1 ||
           0b00 syl2 ||
           0b00 long_op=[0..29] && 0b01 long_op=[30..59] }
  SYNTAX { syl1 || syl2 || long_op ";" }
}

...

OPERATION long_op
{
  DECLARE { ... }
  CODING { opcode_format immediate=0bx[32] }
  SYNTAX { opcode_format "," immediate=#X32 }
}
```

Figure 7.27. LISA description of 32-bit immediate extension.

Considering the whole ST200 architecture, the LISA model contains 141
LISA operations. A total of 5035 lines of LISA code have been written to describe the complete processor. It took two weeks to implement a first instruction
set accurate model without validation. This model reaches a simulation speed
of 8.2 MIPS on a 1.2 GHz Athlon™. It took four weeks to refine this model to
a cycle accurate one.

6.3.4   C Compiler Support.   A C compiler is generated from the ST200
LISA model using the LISATek™ Compiler Designer. As described in Section 4.4.2, the Compiler Designer relies on CoSy®[10] Express from ACE. The
verification of the generated compiler is based on the SuperTest compiler verification suite which is a test and validation suite for C/C++ compilers from
ACE containing more than 400,000 conformance and quality checks [9].

To evaluate the code quality of the generated C compiler, the ST multi-flow
compiler from STMicroelectronics is used as reference, which accomplishes

---

[10]CoSy is an international trademark of ACE Associated Computer Experts bv., *http://www.ace.nl.*

ST200 specific optimizations. Both the compiler development and the realization of optimizations are currently reserved to compiler experts and are very time consuming. In Figure 7.28, the black columns represent code produced by the ST multi/flow compiler, normalized to 100%.

Several DSP algorithms are used as benchmarks. The generated compiler, without architecture-specific optimizations, produces an increase of code size and execution cycle count with a factor of 3.6 at maximum. Due to its automatic generation, however, the development time is dramatically shortened and thus the generated compiler is well suited for design space exploration. Using CoSy®, the generated compiler can be optimized for the ST200 architecture to achieve production quality as soon as the architecture specification is fixed.

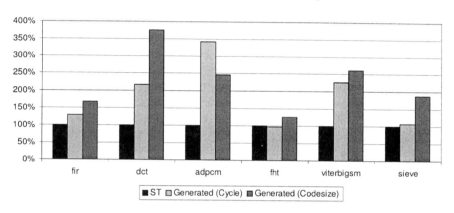

Figure 7.28. Code quality of generated ST200 compiler.

7. Modeling Interconnects of an MPSoC

7.1 Need for Early System Simulation

Market needs and fierce competition in the electronics industry pose enormous challenges to conceptualize, implement, verify, and program today's complex SoC designs. System architects are obliged to employ heterogeneous computational fabrics to meet conflicting requirements with respect to performance, flexibility, and power efficiency. The complexity problem of increased heterogeneity also applies to the on-chip communication. Already today numerous buses of different types are necessary to cope with the escalating data traffic [178]. In the near future full scale Networks-on-Chip will arise that provide the required performance as well as Quality-of-Service (QoS) [22, 200, 216].

In the predominant industrial practice, design of complex and programmable SoCs still follows the traditional flow, where the textual architecture specification phase is followed by sequential and mostly decoupled implementation of hardware and software parts. Finally processors, buses, memories, and dedicated logic blocks are integrated on the fully pin, bit, and cycle-accurate Register Transfer Level (RTL). The high complexity and poor simulation speed on this implementation level prohibits consideration of architectural tradeoffs, and integration of the embedded software has to wait even until the silicon is available. Due to the lack of early system integration, complex SoCs are either over-designed or fail to fulfill the specified performance requirements. Especially, the performance related to the software part, like CPU load, impact of the Real Time Operating System (RTOS), or software response time, can hardly be analyzed without a cycle accurate simulation model.

Furthermore, the software performance is heavily affected by the shared communication and memory architecture. Thus, an isolated consideration of a single processor may hide potential bottlenecks due to bus utilization and memory access latency. Such issues have to be addressed in the design flow as early as possible to prevent late and costly changes of the architecture specification.

7.2   Abstraction Levels: From Timed to Cycle Accurate

Thus, the processor cores in the SoC are not the only components that need to be modeled applying an abstract description language. Also the overall SoC behavior, especially the intermodule communication, needs an efficient abstract modeling style (Table 7.2).

Table 7.2.   Modeling abstraction levels.

communication abstraction	data abstraction	timing abstraction	processor module abstraction
	packets	untimed	functional specification
		timed	performance model
TLM			
	bytes/words	transaction	instr. accurate ISS
		transfer	cycle accurate ISS
RTL	bit-vectors	cycle	HDL processor model

For the inter-module communication, all levels higher than RTL are covered by the new Transaction-Level-Modeling (TLM) paradigm [89, 236]. These levels have in common that the pin wiggling on the module interfaces is not modeled in detail anymore, but is replaced by condensed Interface Method Calls (IMC). A distinction within this large range of TLM can be done with respect to the abstraction level of data and timing. *Byte/word accurate* TLM achieves a significant simulation performance gain against RTL without sacrificing cycle accuracy when modeling at *transfer* timing level. *Packet level* TLM, in contrast, obtains a further high gain in simulation speed especially by modeling a whole burst of data transfers as one single communication event.

7.3 Co-Simulation and System Simulation

Normally, for simulating the entire SoC, models of comparable abstraction levels are combined. As shown in Table 7.2, *transfer* TLM communication models together with a cycle accurate Instruction Set Simulator (ISS) preserve the full cycle accuracy as known from RTL.

Earlier in the design flow, instruction accurate ISSs are typically combined with *transaction* TLM communication models. Both sides are driven by their system clock, but in both cases, some inaccuracies might occur due to the more abstract modeling. On the ISS side, all cycles lost due to the existence of the pipeline cannot be modeled properly since the pipeline is not yet contained on this abstraction level. Similarly, the communication models on *transaction* TLM level perform the transfer of whole bytes or words in one module invocation, without necessarily modeling the parallelism which is possible on the cycle level.

In very early architecture design phases, an ISS is typically not yet available. For most modules, the decision is not even made whether the functionality is to be implemented in hardware or in software. The architectural exploration supporting basic decisions like these is performed using the initially developed functional specifications. The module functionality is modeled abstractly in C/C++ and the estimated time consumption of every computation can be annotated to these models. Similarly, on the communication side, the *packet* based TLM paradigm is applied. In one simulation event, entire data packets like IP packets are transferred at once. As for the modules, the time consumption for the communication process can be annotated.

However, for the overall system simulation during the design process, it is often reasonable to combine models of different abstraction levels. This enables a successive design flow [242]. While refining one system module or communication model to a lower abstraction level, it can directly be verified in the very same system context. Additionally, by integrating the modules currently under consideration into a more abstract environment, high overall simulation speed is achieved while obtaining realistic stimuli. For the early development of the embedded software, no cycle accurate ISS or even RTL model is necessary. For this purpose instruction accurate ISSs are sufficient, and they can be embedded into a faster packet level communication environment for modeling the cache refills over the network [241].

In order to combine different abstraction levels, extended adapters are necessary which do not only connect the modules one-to-one, but also bridge the abstraction gap. For instance, to adapt the timing accuracy, a state machine is contained in the adapter. Such a state machine implements the cycle accurate protocol timing which is not performed by the abstract side itself.

7.4   The SystemC Library

The emerging EDA standard for System Level Design is the SystemC™ library. This C++ class library enables efficient modeling of hardware by providing an event-driven simulation engine as well as support for the TLM abstraction level. In this section the basic concepts of SystemC 2.0.1 based System Level Design are briefly introduced. A thorough representation of this topic area is given by Groetker et al. [89].

As illustrated in Figure 7.29, SystemC 2.0.1 follows the *Interface Method Call (IMC)* principle to achieve high modularity in communication modeling. Processes are wrapped into modules and access communication services through ports. The available methods are declared in the interface specification and implemented by the channel. Graphically a port is represented as a square containing two opposing arrows and an interface is represented as a circle containing a U-turn arrow. A channel implementing an interface is a hexagon.

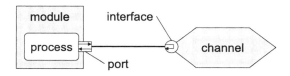

Figure 7.29.   Interface Method Call principle.

8.   The NoC Exploration Platform GRACE++

Future MPSoC designs will contain dozens or even hundreds of system modules, and the resulting communication issues will be one of the limiting system bottlenecks. Thus, bus topologies will no longer be the best suited global communication infrastructures [22, 216]. Instead, more general network topologies will be implemented on the chips (Network-on-Chip, NoC)[6, 224] and these need to be designed carefully.

An efficient design space exploration is only possible if the communication architecture models are decoupled from the processing elements and the application. Only with this approach of orthogonalization, different communication models and topologies can be set up quickly without modifying the system modules themselves.

On the high packet TLM abstraction level, both are possible: A high modeling efficiency due to the orthogonalization of communication and computation modeling, as well as a high simulation speed because of the high abstraction level.

This section introduces the simulation framework GRACE++ for system level exploration of the communication architecture [129]. It has been commercialized as "Architects View Framework" (AVF) by CoWare, Inc[11]. The basic design principles and the underlying models of computation are based on the seminal work of Sangiovanni-Vincentelli et al. [144, 204]. First, this section presents a unified approach to the modeling of on-chip communication and the associated NoC architecture exploration methodology. Afterwards, the internal structure of the framework and the algorithms are elaborated, which capture the performance characteristics of the distinctive on-chip networks.

8.1 Overview

The GRACE++ NoC framework enables the systematic design space exploration of complex on-chip communication networks. The key idea of the NoC framework is to simplify the time-consuming process of changing the communication architecture and topology. This is achieved by the following two concepts:

- *Generic Synchronization Interface:* The communication architecture is hidden behind the already introduced generic synchronization interface. This attains the complete orthogonalization of the synchronization services employed by the functional processes from the communication architecture. For instance, the communication architecture can change from point-to-point towards a bus-centric architecture, without the modification of either the functional processes or the interface.

- *Descriptive Instantiation:* All topology aspects and all parameter options of the communication architecture are specified through configuration files that are elaborated during the initialization phase of the simulation. Thus the whole instantiation and binding of the communication architecture is automatically done by the NoC framework. Only some parameters need to be changed in the configuration file for simulating alternative architectures. Additionally, no recompilation is needed to iterate over different communication alternatives, which dramatically speeds up design space iteration.

Figure 7.30 illustrates the transparent mapping of the inter-module communication onto two alternative bus architectures. Thanks to the generic synchronization interface and the descriptive instantiation mechanism, the transition

---

[11] *http://www.coware.com*

from one bus topology to another is performed by a modification of the config-uration files.

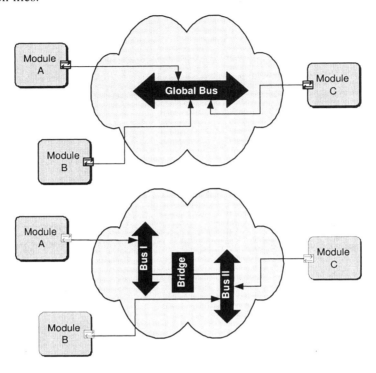

Figure 7.30. Transparent communication mapping.

As a major differentiator against general performance measurement frame-works [18, 253, 75], the NoC framework covers a broader space than only contemporary point-to-point and shared bus architectures. Instead, the NoC framework also addresses emerging Network-on-Chip architectures, which are generally believed to provide the global on-chip communication infrastructure of future multi-processor platforms. Thus, the proposed MPSoC workbench and exploration methodology are designed to sustain emerging architecture trends.

8.2 NoC Exploration Methodology

The GRACE++ NoC framework enables the systematic design space explo-ration of complex on-chip communication networks. During the simulation, the evaluation modules connected to the architecture specific network engines col-lect statistical information about resource utilization, latency, and throughput.

Based on this data, the system architect designs the communication infrastructure according to the following successive refinement steps:

- *Initial Throughput Measurement:* During the initial throughput measurement, the overall on-chip traffic is functionally captured by means of an unconstrained point-to-point network engine. The resulting communication profile identifies interacting partners and rough throughput requirements.

- *Coarse Network Partitioning:* The coarse network partitioning is dedicated to the identification of the optimum mix of network types. The system architect maps the point-to-point communication to an appropriate set of network types by configuring the generic part of the simulation framework with the corresponding set of network-specific engines.

- *Parameter Calibration:* By iterative parameter calibration, the selected communication architecture is fine-tuned to the traffic requirements. Parameters refer to the bandwidth of a bus system or the queue length of a crossbar architecture.

Of course the network partitioning has key impact on the final quality of results. Here the unified approach enables a rapid exploration of different network architectures by simply exchanging the network engines. By that the system architect can optimize the communication in an iterative exploration cycle. Architecture debugging as well as design decisions are supported by a multitude of visualization techniques (Figure 7.31). In case of very complex applications, the simulation driven approach can be complemented with statistical post-processing techniques like regression analysis to reduce the overall design space.

8.3   Unified Modeling of On-Chip Communication

In general, two different kinds of modules can participate in a communication event or *transaction*: While master modules actively initiate transactions, the slave modules can only react passively. Typical masters are processors, DMA controllers, or autonomous ASIC blocks, whereas typical slaves are memories or co-processors. Of course peer-to-peer communication between two masters is also possible.

This general master-slave communication scheme is reflected in the overall organization of the NoC evaluation framework depicted in Figure 7.32.

On-chip communication services are offered through a generalized master interface. In compliance with the *Interface Based Design* principle [204] [42], master modules initially not compliant to this interface can be attached by means of adapters.

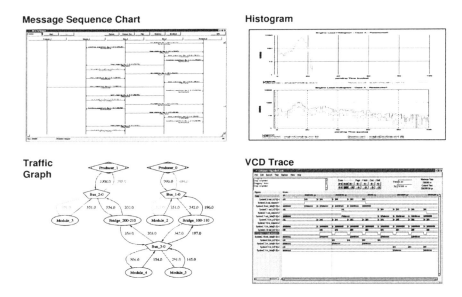

Figure 7.31.   NoC debugging and analysis.

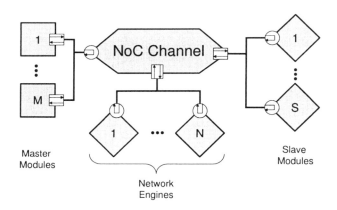

Figure 7.32.   NoC channel overview.

The central module of the NoC simulation framework is represented by the *NoC channel*, which captures the generic part of the communication function-ality. This comprises, e.g., the implementation of the synchronization protocol, the general transaction handling, and keeping track of the user module status.

The attached *network engines* are responsible for modeling the communica-tion architecture specific information, especially the interconnect topology.

8.4 Network Engine Algorithms

The properties of the communication architecture are defined by selecting and parameterizing a suitable network engine. The NoC framework offers three generic engines for the main network types:

- *Point-to-Point Engine:* The point-to-point engine represents an exclusive resource between a single initiator and a single target. The engine is configured by a bandwidth matrix which defines the bit width for every path between the M masters and the $M + S$ possible destination modules. Incoming requests are buffered in a FIFO queue as long as the respective resource is occupied.

- *Bus Engine:* The bus engine models a shared resource between multiple initiator and target modules. Pending requests are queued until the access to the bus is granted by the arbitration process. The engine can be specialized towards priority or TDMA-like bus arbitration schemes by different implementations of the arbitration algorithm. The priority based bus arbiter simply selects the request with highest priority, whereas the TDMA arbiter is based on a static allocation table. The bus engine is configured by the bandwidth parameter to determine the duration of the actual transaction, which starts after a request has been granted.

- *Crossbar Engine:* The crossbar engine embodies router-like parallel communication resources with a centralized arbitration scheme. As an example, the NoC library contains an engine which models Virtual Output Queued (VOQ) [230] architectures for equal-size data packets.

A *Hierarchical Engine* enables the composition of basic engines to complex network topologies. Additionally, the designer can implement his own network engines to model any communication architecture he wants to explore. This way, network engines have been developed which model existing communication infrastructures like STBus[12], ARM AMBA[TM] AHB[13] and Sonics SiliconBackplane[TM] III [224].

9. Summary

In this chapter, the LISA and GRACE++ tools are presented. Supporting a very similar design methodology to the MESCAL approach, the alternative tools enable a wider design space.

---

[12]STBus, ST Microelectronics, *http://www.st.com.*
[13]ARM AMBA, Advanced RISC Machines (ARM), *http://www.arm.com.*

We fully agree with the MESCAL approach that programmable platforms are the only way of meeting today's and tomorrow's flexibility constraints, and tailoring these specifically for the target application domain is the key to meet the performance demands with a good energy efficiency. Designing these platforms requires a systematic methodology, which again needs suitable tooling for obtaining optimal results in a reasonable design time.

The LISA processor design environment supports such a systematic methodology by opening a huge design space to find an optimal processor architecture for a given application. Since the LISA tools generate the hardware implementation as well as the software development tools including C compiler automatically, the designer actually can cope with, and even exploit, the large design space.

On multiple levels of abstraction, in all phases of the processor design flow, the LISA processor models can be integrated into a suitable system model. The GRACE++ tools allow modeling the overall MPSoC on the highest possible abstraction level which takes architectural properties into account. On this level efficient design space exploration is possible to find an optimal on-chip communication architecture for interconnecting the tens or hundreds of system modules of modern MPSoCs.

LISA and GRACE++ have been developed at the Aachen University of Technology and are commercialized by CoWare. The product names are LISATek™ and AVF (Architects View Framework).

Chapter 8

COMMERCIAL CONFIGURABLE PROCESSORS AND THE MESCAL APPROACH

Grant Martin and Steve Leibson
Tensilica, Inc.
Santa Clara, CA

In this chapter, we will discuss how commercial configurable and extensible processors offered by Tensilica in its V and LX series, fit into the MESCAL design and methodology approach for ASIPs and programmable platforms. We will highlight a number of ways in which the Tensilica ASIP design approach adopts and further extends the MESCAL methodology. As the authors of Part I note, their experience in working with the Tensilica Xtensa® processor design approach both predates the MESCAL methodology and was indeed one of the key influences that led them to codify the methodology.

To review the MESCAL approach, let us summarize its five basic elements from Chapter 1:

1 *Judiciously Using Benchmarking* – ASIPs should be application driven, with careful attention paid to benchmarks driving the design. They should be representative of the targeted application domain, easy to specify and manageable, and should enable a quantitative comparison of the architectures being considered.

2 *Inclusively Identifying the Architectural Space* – careful definition of the appropriate design space is important for effective design space exploration. Special functions should be constrained to those suitable to the application, and perhaps to those drawn from libraries.

3 *Efficiently Describing and Evaluating the ASIPs* – software tools to map application benchmarks onto each design point must be available and they must allow the quality of the mapping to be measured. A retargetable software environment is necessary, preferably driven by a central archi-

tecture description to ensure consistency across the compiler, functional, timing and power simulators, and HDL generation.

4 *Comprehensively Exploring the Design Space* – this should be a feedback driven process, where each design point is evaluated on the appropriate metrics, using the software environment described in element 3. Efficient exploration for a large number of design points relies on quick reconfiguration of the design environment and simulators.

5 *Successfully Deploying the ASIP* – ASIPs that have been created to be a good match between application and architecture must be able to be programmed efficiently at a reasonable level, using programmer's models that expose critical features of the ASIP architecture and enable exploitation of unique architectural features. At the same time, application portability is an important goal.

In this chapter, we will first describe the Tensilica approach to ASIP design and follow that by an analysis of how well this approach follows the MESCAL approach. We will conclude with some thoughts about how the Tensilica ASIP design methodology is being expanded to include multiprocessor design considerations (also known as MPSoC) and again compare that to the MESCAL methodology.

1.   The Tensilica Approach to Application-Specific Instruction-Set Processors

The Tensilica approach to configurable, extensible processor design, derived from a base instruction-set architecture (ISA) using automated generation processes, has been well-documented over the years, both in the early days [82] and quite recently [203]. Nevertheless, in order to better understand how it relates to the MESCAL approach, it is useful to start with an overview of the range of configurability possible with Tensilica's processors.

Tensilica processors are based on two micro-architectures. The initial processor micro-architecture, which has evolved from the original Xtensa® processor, in Xtensa® II to V generations, is based on a five-stage single-pipeline RISC with a number of configuration options. Instruction extensions are described in an ISA description language called TIE, and are compiled deep into the processor's datapath. We can regard the configuration/extension options as being "coarse-grained" (structural) and "fine-grained" (instruction extensions via TIE). For the Xtensa® processors, the configuration options include:

- Register-file size.

- Specialized functional units, including 16- and 32-bit multipliers, multiplier-accumulator (MAC), floating-point unit, DSP instruction unit.

- On-chip debug, tracing, and JTAG ports.

- A wide variety of local memory interfaces including instruction and data caching, local instruction and data RAM and ROM, general-purpose local memory interface (XLMI), and memory management capabilities including protection and translation. These interfaces are configurable as to size and address space mapping.

- Timers, interrupts, and exception vectors.

- Processor bus interface (PIF), including width, protocol, and address decoding.

- Ability to link system RAM and ROM to the processor bus interface (PIF).

- TIE instruction extensions included special purpose functions, dedicated registers, wire interfaces mapped into instructions.

The second-generation Tensilica micro-architecture, introduced in mid-2004 and known as the Xtensa® LX generation [117], offers a number of significant additional options for configurability and instruction extension:

- Multi-issue architecture with an opportunity for multiple execution units to be active simultaneously, configured by the user.

- Flexible-length instruction-set encoding (FLIX™) for efficient code size and an ability to intermix instruction widths and create the multi-operation instructions to control the multi-issue architecture.

- Dual load/store option to increase the classical ISA bandwidth.

- Either five- or seven-stage pipeline, the latter to improve the matching between processor performance and memory speed.

- New options for communications directly into and out of the processor's execution units, via TIE ports and queues. These provide an ability to hook FIFO communications channels (either unbuffered, or n-deep) directly into instruction extensions defined in the TIE language. As a result, multi-processor systems using direct FIFO-based communications between execution units in the processor datapath are now possible.

TIE, the Tensilica Instruction Extension language, is a Verilog®-like specification language that allows one to define instruction extensions that are directly implemented as hardware in the processor datapath and are then accessible to software control in user code. The TIE language allows one to specify the input and output arguments of the instructions, the storage elements/registers used by them, and the syntax and semantics of the instruction (defining both is important for verification processes during processor generation). Tensilica's TIE compiler transforms these instruction extensions into both hardware implementations and modifications to the software tool flow so that compilers, debuggers, and related SW tools are aware of them.

As described in [203], the addition of these extra dimensions of configurability and extensibility significantly opens up the design space that is possible to explore with Tensilica embedded processors. Beyond the base ISA, the processor can be configured with multiple execution units. With clever compilation and instruction encoding, these multiple execution units can be controlled by embedded software in an efficient and highly application-specific way. Multiple configured processors can be assembled into a subsystem using a variety of inter-task communications semantics to accomplish an embedded application. In particular, the ability to hook FIFO queues directly into the execution pipeline so that data produced by a task running on one processor can be directly fed into a task running on a second, opens up the possibility of building high performance subsystems that can substitute for hardware implementations of functions in many application areas. Yet, being based on an ISA approach, these systems can be modified in their in-field function using standard software methods.

To complement the configurability and extensibility of the processor architecture, it is important to have a highly automated processor generation-process to speed development and reduce design errors. In Tensilica's case, this is offered via a web-based configuration service accessible to designers via client-based software. The automated process allows designers to enter configuration parameters via a series of entry screens, where the feasibility of the set of configuration parameters is cross-checked during entry. The designer is given warnings or errors when an erroneous combination of parameter settings is chosen. Instruction extensions are created in the TIE language using an editor that is part of the Tensilica Integrated Development Environment, Xtensa Xplorer^{TM}, which is an IDE based on the Eclipse open-source framework. The TIE compiler checks the designer-defined instruction extension code and generates a local copy of the relevant software tools (compiler, instruction-set simulator), which can be used for rapid software development and experimentation. Instruction extensions can be automatically generated using a tool called XPRES^{TM}, which is described later in this chapter.

At some point, the designer will want to generate the processor configuration, the complete set of software tools, and potentially the HDL hardware description. The generation process allows the configuration files and designer-specified TIE files to be uploaded to the Tensilica secure configuration server environment, and a variety of configuration runs (e.g. software-only, to generate tools, or software-hardware, which will also generate an RTL implementation for the configuration) can be initiated. Within one to two hours, the complete configuration will be generated, verified, and the tools and configuration outputs can be downloaded to a local workstation for further work on software development, instruction extension, and performance analysis and optimization. This automated configuration process is extensively verified [116] in the process of developing each new release of Tensilica software and hardware, and has been validated in field use by more than 70 licensees on many different processor instantiations over the last several years.

This brief overview of the Tensilica processor architecture and configuration and extension possibilities is elaborated in more detail in [203]. We will now describe how the design approach using this commercial processor generator fits into the MESCAL five-element methodology.

2. Element 1: Judiciously Using Benchmarking

We summarized the first element of the MESCAL approach, judiciously using benchmarking, as "ASIPs should be application driven, with careful attention paid to benchmarks driving the design. They should be representative of the targeted application domain, easy to specify and manageable, and should enable a quantitative comparison of the architectures being considered."

Tensilica employs two basic approaches to benchmarking, motivated by the fact that any configuration of the Xtensa® processor is almost certainly different from any other, and that these are drawn from a near-infinite number of possible configured and extended processors in the configuration space offered by the processor architecture and automated generation process.

The first part of this approach has been to help designers decide whether an Xtensa® processor is appropriate for their application through the use of standard industrial application-oriented benchmarks. These can give a good indication of whether both the inherent performance of the processor architecture, and the range of improvements made possible through instruction extension, will make it a reasonable choice for an application which lies in the same domain as the standard benchmarks.

The second part of this approach provides the automated design flow, software tools, and instruction-set simulators that allow application code to be directly measured on a variety of configurations of the Xtensa® processor. To the extent that a reasonable partial or near-complete implementation of the ultimate application code is available, this code is the best benchmark both for the base processor capabilities and to guide the instruction extension process.

2.1 Standard Industry Benchmarks

There are many processor benchmarks in the industry – an excellent history and overview of their evolution is found in [147]. Two sets are most relevant to a wide variety of embedded systems design and choice of processing cores: BDTI and EEMBC.

Berkeley Design Technology, Inc. (BDTI) is a technical services company that has focused exclusively on the applications of DSP processors since 1991. BDTI serves as a private third party that develops and administers DSP benchmarks. In 1994, BDTI introduced its core suite of DSP benchmarks, consisting of 12 algorithm kernels that represent key operations used in common DSP applications. These were revised in 1999. The 12 DSP kernels in the BDTI Benchmark appear in Table 8.1.

The BDTI Benchmark is a hybrid benchmark that uses code (kernels) extracted from actual DSP applications. BDTI calls its benchmarking approach "algorithm kernel benchmarking and application profiling." The algorithm kernels used in the BDTI benchmarks are functions that constitute the building blocks used by most signal-processing applications, and are the most computationally intensive portions of the donor DSP applications.

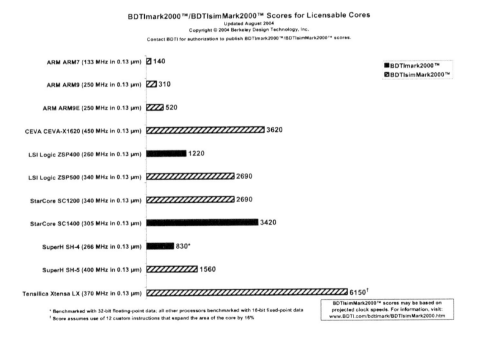

Figure 8.1. The BDTImark2000™/BDTIsimMark2000™. Scores © 2004 BDTI.

To satisfy people who use one number to make a rough cut of processor candidates, BDTI rolls a processor's execution times on the 12 DSP kernels into one composite number that it has dubbed the BDTImark2000. To differentiate

Table 8.1.   BDTI benchmark kernels.

Kernel Function	Function Description	Example Applications
Real block FIR	Finite impulse response filter that operates on a block of real (not complex) data.	Speech processing (e.g. G.728 speech coding)
Complex block FIR	FIR filter that operates on a block of complex data.	Modem channel equalization
Real single-sample FIR	FIR filter that operates on a single sample of real data.	Speech processing, general filtering
LMS adaptive FIR	Least-mean-square adaptive filter, operates on a single sample of real data.	Channel equalization, servo control, linear predictive coding
IIR	Infinite impulse response filter that operates on a single sample of real data.	Audio processing, general filtering
Vector dot product	Sum of the point-wise multiplication of two vectors.	Convolution, correlation, matrix multiplication, multi-dimensional signal processing
Vector add	Point-wise addition of two vectors, producing a third vector.	Graphics, combining audio signals or images
Vector maximum	Find the value and location of the maximum value in a vector.	Error control coding, algorithms using block floating-point
Viterbi decoder	Decode a block of bits that have been convolutionally encoded.	Error control coding
Control	A sequence of control operations (test, branch, push, pop, and bit manipulation).	Virtually all DSP applications include some control code
256-point, in-place FFT	Fast Fourier Transform converts a time-domain signal to the frequency domain.	Radar, sonar, MPEG audio compression, spectral analysis
Bit unpack	Unpacks variable-length data from a bit stream.	Audio decompression, protocol handling

processor benchmark scores verified with processor chips from simulated processor core scores, BDTI publishes the results of simulated benchmark scores (for processor cores) under a different name: The BDTIsimMark2000. Both the BDTImark2000 and the BDTIsimMark2000 scores are available without charge on BDTI's Web site[1]. Figure 8.1 shows BDTImark2000™ and BDTIsim-Mark2000™ scores for ARM's ARM7® and ARM9E®, CEVA CEVA-X™, LAS LSI Logic ZSP400™ and ZSP500™ StarCore™ SC1200 and SC1400, SuperH SH-4 and SH-5, and Tensilica Xtensa® LX microprocessor and DSP cores.

The BDTI benchmarks are a starting point for evaluating processor cores, but in meeting the benchmark criteria discussed earlier in this book (see Chapter 2), they are far from a complete benchmarking approach. They suffer both from being based on only kernels, and from not being specific to a particular application for which a processor selection is being made. Clearly, to make a proper processor selection, it is necessary to go much further.

The second standard industrial set of benchmarks is from EEMBC [61], a non-profit, embedded-benchmarking organization (EDN Embedded Benchmark Consortium) started in 1997. EEMBC's stated goal was to produce accurate and reliable metrics based on real-world embedded applications for evaluating embedded processor performance. EEMBC (pronounced "embassy") spent nearly three years developing a suite of benchmarks for testing embedded microprocessors and introduced its first benchmark suite at the Embedded Processor Forum in 1999. EEMBC released its first certified scores in 2000 and, during the same year, announced that it would start to certify benchmarks run on simulators so that processor cores could be benchmarked in addition to processor chips. As of 2004, EEMBC had more than 50 corporate members.

The EEMBC benchmarks are contained in six suites loosely grouped according to application. The six suites are: Automotive/industrial, consumer, Java GrinderBench, networking, office automation, and telecom. Each of the suites contains several benchmark programs that are based on small, derived kernels extracted from application code. All EEMBC benchmarks are written in C, except for the benchmark programs in the Java benchmark suite. The six benchmark suites and descriptions of the programs in each suite appear in Tables 8.2 through 8.7.

EEMBC's benchmark suites with their individual benchmark programs allow designers to select the benchmarks that are relevant to a specific design, rather than lumping all of the benchmark results into one number. EEMBC's benchmark suites are developed by separate subcommittees, each working on one application segment. Each subcommittee selects candidate applications that represent the application segment and dissects each application for the key kernel code that performs the important work. This kernel code coupled with a test harness becomes the benchmark. Each benchmark has published

---

[1] *http://www.bdti.com*

Table 8.2. EEMBC Automotive/Industrial benchmark programs.

Benchmark Name	Benchmark Description	Example Application
Angle to Time Conversion	Compute engine speed and crankshaft angle	Automotive engine control
Basic Integer and Floating Point	Calculate arctangent from telescoping series	General purpose
Bit Manipulation	Character and pixel manipulation	Display control
Cache "Buster"	Long sections of control algorithms with pointer manipulation	General purpose
CAN Remote Data Request	Controller Area Network (CAN) communications	Automotive networking
Fast Fourier Transform (FFT)	Radix-2 decimation in frequency, power spectrum calculation	DSP
Finite Impulse Response (FIR) Filter	High- and low-pass FIR filters	DSP
Inverse Discrete Cosine Transform (iDCT)	iDCT using 64-bit integer arithmetic	Digital video, graphics, image recognition
Inverse Fast Fourier Transform (iFFT)	Inverse FFT performed on real and imaginary values	DSP
Infinite Impulse Response (IIR) Filter	Direct-for II, N-cascaded, second-order IIR filter	DSP
Matrix Arithmetic	LU decomposition of NxN input matrices, determinant computation, cross product	General purpose
Pointer Chasing	Search of a large, doubly linked list	General purpose
Pulse Width Modulation (PWM)	Generate PWM signal for stepper-motor control	Automotive actuator control
Road Speed Calculation	Determine road speed from successive timer/counter values	Automotive cruise control
Table Lookup and Interpolation	Derive function values from sparse 2-D and 3-D tables	Automotive engine control, anti-lock brakes
Tooth to Spark	Fuel-injection and ignition-timing calculations	Automotive engine control

Table 8.3.   EEMBC consumer benchmark programs.

Benchmark Name	Benchmark Description	Example Applications
High Pass Grey-Scale Filter	2-D array manipulation and matrix arithmetic	CCD and CMOS sensor signal processing
JPEG	JPEG image compression and decompression	Still-image processing
RGB to CMYK Conversion	Color-space conversion at 8 bits/pixel	Color printing
RGB to YIQ Conversion	Color-space conversion at 8 bits/pixel	NTSC video encoding

Table 8.4.   EEMBC GrinderBench Java 2 Micro Edition (J2ME) benchmark programs.

Benchmark Name	Benchmark Description	Example Applications
Chess	Machine-vs.-machine chess matches, 3 games, 10 moves/game	Cell phones, PDAs
Cryptography	DES, DESede, IDEA, Blowfish, Twofish encryption and decryption	Data security
kXML	XML parsing, DOM tree manipulation	Document processing
ParallelBench	Multithreaded operations with mergesort and matrix multiplication	General purpose
PNG Decoding	PNG image decoding	Graphics, Web browsing
Regular Expression	Pattern matching and file I/O	General purpose

Table 8.5.   EEMBC Networking benchmark programs.

Benchmark Name	Benchmark Description	Example Applications
IP Packet Check	IP header validation, checksum calculation, logical comparisons	Network router, switch
IP Network Address Translator (NAT)	Network-to-network address translation	Network router, switch
OSPF version 2	Open shortest path first/Djikstra shortest path first algorithm	Network routing
QoS	Quality of service network bandwidth management	Network traffic flow control

Table 8.6. EEMBC Office Automation benchmark programs.

Benchmark Name	Benchmark Description	Example Applications
Dithering	Grayscale to binary image conversion	Color and monochrome printing
Image Rotation	90-degree image rotation	Color and monochrome printing
Text Processing	Parsing of an interpretive printing control language	Color and monochrome printing

Table 8.7. EEMBC Telecom benchmark programs.

Benchmark Name	Benchmark Description	Example Applications
Autocorrelation	Fixed-point autocorrelation of a finite-length input sequence	Speech compression and recognition, channel and sequence estimation
Bit Allocation	Bit-allocation algorithm for DSL modems using DMT	DSL modem
Convolutional Encoder	Generic convolutional coding algorithm	Forward error correction
Fast Fourier Transform (FFT)	Decimation in time, 256-point FFT using Butterfly technique	Mobile phone
Viterbi Decoder	IS-136 channel decoding using Viterbi algorithm	Mobile phone

benchmarking guidelines in an attempt to force the processor vendors to play fair.

EEMBC employs EEMBC Certification Laboratories, LLC (ECL) to conduct benchmark tests and enforce the rules of EEMBC benchmarking. Although the processor vendor ports the EEMBC benchmark code and test harnesses to its own processor and runs the initial benchmark tests, only ECL can certify the results. ECL takes the vendor-supplied code, test harness, and a description of the test procedures used and then certifies the benchmark results by inspecting the supplied code and re-running the tests. In addition to code inspection, ECL makes changes to the benchmark code (such as changing variable names) to counteract "over optimized" compilers that substitute hand-optimized code for key portions of the benchmark. Vendors cannot publish EEMBC benchmark scores without ECL certification.

EEMBC rules allow for two levels of benchmarking. The lower level of EEMBC produces "out-of-the-box" scores. Out-of-the-box EEMBC benchmark tests can use any compiler (in practice, the compiler selected has changed the performance results by as much as 40%) and any selection of compiler switches, but cannot modify the benchmark source code. The "out-of-the-box" results therefore give a fair representation of the abilities of the processor/compiler combination without adding programmer creativity as a wild card. The higher level of EEMBC play is called "full-fury." Processor vendors seeking to improve their full-fury EEMBC scores (posted as "optimized" scores on the EEMBC Web site) can use hand-tuned code, assembly-language subroutines, special libraries, special CPU instructions, coprocessors, and other hardware accelerators. Full-fury scores tend to be much better than out-of-the-box scores, just as application-optimized production code generally runs much faster than code that has merely been run through a compiler.

EEMBC's work is ongoing. To prevent benchmark rot, EEMBC's subcommittees constantly evaluate revisions to the benchmark suites. The Networking suite is already on version 2.0 and the Consumer suite is undergoing revision as of the writing of this chapter. To date, EEMBC has created benchmarks that focus on measuring execution speed but the organization is developing power- and energy-oriented benchmarks because power efficiency and low-power operation have become important processor characteristics. These power and energy benchmarks should become available from EEMBC in 2005.

The ability to tailor a processor's ISA to a target application (which includes benchmarks) can drastically increase execution speed of that program. For example, Figure 8.2 shows EEMBC Consumer benchmark suite scores for Tensilica's configurable Xtensa® V microprocessor core (labeled Xtensa® T1050 in Figure 8.2). The right-hand column of Figure 8.2 shows benchmark scores for a standard processor configuration. The center column shows the improved benchmark scores for a version of the Xtensa® V core that has been tailored by a design engineer specifically to run the EEMBC Consumer benchmark programs. Table 8.8 summarizes the performance results of the benchmarks for the stock and tailored processor configurations.

Note that the stock Xtensa® V processor core performs on par with or slightly faster than other 32-bit RISC processor cores. Tailoring the Xtensa® V processor core for the individual tasks in the EEMBC Consumer benchmark suite by adding new instructions that are matched to the needs of the benchmark computations boosts the processor's performance on the individual benchmark programs from 4.6x to more than 100x depending on the benchmark. The programmer modifies the benchmark or target application code by adding C intrinsics that use these processor extensions. The resulting performance improvement is similar to results that designers might expect to achieve when tailoring a processor for their target application code so in a sense, the EEMBC benchmarks are still performing exactly as intended. The optimized Xtensa® V EEMBC Consumer benchmark scores reflect the effects of focusing three

Consumer Benchmarks		
Processor Name-Clock	Tensilica Xtensa T1050-260	Tensilica Xtensa T1050-260
ConsumerMark	2.02256 (at 1 Mhz)	.08696 (at 1 Mhz)
Vendor Score Interpretation	**View Interpretation**	**View Interpretation**
Type of Platform	Simulator	Simulator
Type of Certification	Optimized	Out-of-the-box
Certification Date	8/1/2002	7/27/2002
Certified by	$\overset{...}{\text{s}}\text{CL}$	$\overset{...}{\text{s}}\text{CL}$
Benchmark Notes	Optimized using 'C' only, no assembly.	
Simulator Type	high level cycle accurate	high level cycle accurate
Native Data Type	32	32
Architecture Type	Configurable, Extensible RISC	RISC
L1 Instruction Cache Size (kbyte)	16	16
L1 Data Cache Size (kbyte)	16	16
External Data Bus Width (Bits)	128	128
Memory Configuration	6-1-1-1 for instructions, 7-1-1-1 for data	6-1-1-1 for instructions, 7-1-1-1 for data
L2 Cache Size (kbyte)	0	0
L2 Cache Clock	none	none
Portability Flags		
Endian	little	little
Warning Flags		
Error Handling and Level		
Debug Settings		
ANSI Adherance		
Include Files		
Code Generation Flags	-O3 -IPA	-O3 -IPA
Post-processor		

Compiler Model and Version	Xtensa C Compiler T1050	Xtensa C Compiler T1050
Floating Point	software	software

Certification Report	**View Report**	**View Report**
Comparison Reports	Compare the processor selected below using EEMBC's **QuickCompare** feature Tensilica Xtensa T1050-260 ⊙	Tensilica Xtensa T1050-260 ⊙

Benchmark Scores	Iterations *per million cycles*	Code Size *in bytes*	Iterations *per million cycles*	Code Size *in bytes*
Compress JPEG	0.22	28,396	.04796	26,896
Decompress JPEG	0.32	29,016	.06837	28,180
High Pass Grey-scale filter	26.98	896	0.45136	900
RGB to CYMK Conversion	26.01	680	0.50251	557
RGB to YIQ Conversion	34.67	660	0.33830	616

Figure 8.2. Tailoring a processor core's architecture for a specific set of programs can result in significant performance gains. © EEMBC. Reproduced by permission.

months worth of a engineering graduate student's "full fury" to improve the Xtensa® V processor's benchmark performance optimizing both the processor and the benchmark code.

The EEMBC benchmarks have a wider application scope than the BDTI kernels and provide more specific design-domain information. Nevertheless, they are still not adequate to select a processor for a specific application. Participation by embedded processor companies in these industrial benchmarking

Table 8.8.   Optimized versus out-of-the-box EEMBC Consumer benchmark scores.

Benchmark Name	Optimized Xtensa® T1050 Score (Iterations/M cycles)	Out-of-the-Box Xtensa® T1050 Score (Iterations/M cycles)	Performance Improvement
Compress JPEG	0.22	0.04796	4.59x
Decompress JPEG	0.32	0.06837	4.68x
High Pass Grey-scale Filter	26.98	0.45136	59.77x
RGB to CMYK Conversion	26.01	0.50251	51.76x
RGB to YIQ Conversion	34.67	0.33830	102.48x

efforts is useful, but not sufficient. The best way to benchmark a processor for a specific design application is to benchmark the actual application code on the processor.

2.2   Application-Specific Benchmarking

Nothing can substitute for benchmarking the actual embedded application of interest on the candidate processors. Of course, if one had to wait for the arrival of the actual hardware, or do all accurate benchmarking on a hardware (RTL) description of the processor, it would be impossible to do any effective benchmarking at all. By providing an automated configuration process that delivers a complete software and simulation environment for any particular configuration within an hour or two, Tensilica makes it possible to run application-specific benchmarks on various configurations of the processor without any particular effort. The Xplorer™ IDE provided as part of the toolset has benchmarking capabilities that allow various processor configurations applied to the same application code to be compared easily for a number of different processor evaluation criteria including total cycle count, memory accesses, cache misses, and so on. All this capability rests on the availability of a cycle-accurate Instruction Set Simulator (ISS), for any particular processor configuration with instruction extensions. Tensilica provides a cycle-accurate ISS for both single-processor and multi-processor simulation and evaluation, created automatically during the configuration process and validated extensively for cycle-accuracy during development.

3. Element 2: Inclusively Identifying the Architectural Space

Element 2 of the MESCAL methodology includes in its definition the "careful definition of the appropriate design space is important for effective design space exploration. Special functions should be constrained to those suitable to the application, and perhaps to those drawn from libraries."

A configurable, extensible processor ISA such as Xtensa® defines a configuration space that is somewhat more constrained than the approach discussed earlier in this book, because it is based on a particular micro-architectural approach. However, this configuration space is still much wider than the design space that is most appropriate for a particular application. The subset of the configuration space that should be explored in detail is found through configuration experiments and subsequent analysis. Alternatively, automated technology that is discussed later can define the appropriate subset. The tradeoff for being a little more constrained is a very high degree of software automation and high-speed search of the more restricted design space. In addition, the concept of special functions being constrained to those suitable to the application is inherent in the idea of adding extensions to the ISA. If instruction extensions are designer-defined and are properly evaluated and chosen using application-specific profiling and benchmarking, then they will inherently be drawn from a set suitable for the application. If instruction extensions are defined by an automated tool driven by application code, as with Tensilica's XPRES™ compiler, then they will very likely be even more constrained to those most suitable for the application. Finally, third-party instruction extensions created by Tensilica application partners in domains such as audio and video processing are the equivalent of application-specific libraries and provide additional choices to designers who want to configure a processor to a design space more rapidly.

There are other aspects to Tensilica processor configuration and options that may be more appropriate to some design applications than others. These include:

- Special-purpose hardware.

- Memory architectures.

- On-chip communications.

- Peripherals.

For example, computationally intensive applications in signal and image processing may be best implemented on processor configurations including floating-point capabilities, DSP-style capabilities such as zero-overhead looping, vector (SIMD) style processing as offered by Xtensa® LX FLIX™ instructions and multi-operation hardware, and hardware multiplier and MAC units. Other applications, such as branch-dominated control algorithms for packet-header classification and processing in networking applications, may not be able to exploit this kind of special purpose hardware at all, leading to area, power, and performance penalties if the inclusion or exclusion of such hardware is not configurable.

Similarly, special memory architectures such as dual X,Y memories (as found in DSP's) can be implemented in Xtensa® LX processors using dual load-store units, and are often most suitable to computationally intensive signal-processing applications, and not necessary for control-centric applications. Image processing, which often involves access to large frame-oriented data with regular access patterns, may be best implemented using processors interfaced to DMA engines and large backing stores. The ability to configure a wide variety of memory interfaces, for both instruction and data storage, with both local and system-accessed memories, and using asynchronous local-memory interfaces and synchronous bus-based interfaces, is needed to match the processor configuration to the application space and is inherent in Tensilica architectures.

Specialized communication facilities, such as the Xtensa® LX processor's TIE ports and queues, allow single or multiple Tensilica processors to be ganged together or linked with extremely low-overhead FIFO channels directly tied to the execution units. Certain types of applications, such as dataflow-style signal and image processing, are a better place to exploit such specialized on-chip communications capabilities; control-dominated applications may be better matched to more orthodox shared-bus and arbitrated-communications architectures.

A wide variety of peripherals are needed for application-specific-processor-based SoC design, but it is difficult for a processor IP provider to anticipate every peripheral and provide it. It is more useful to support a wide variety of interfaces to the processor to allow easy connection to the right peripherals. The Tensilica architectures support both a general-purpose, asynchronous, local memory interface (XLMI), which can be used for high-performance peripherals, and the synchronous processor interface (PIF), which can be attached to any on-chip bus architecture for connection both to low-speed peripherals and high-performance buses and devices.

4. Element 3: Efficiently Describing and Evaluating the ASIPs

Element 3 of the MESCAL methodology implies that "software tools to map application benchmarks onto each design point must be available and they must allow the quality of the mapping to be measured. A retargetable software environment is necessary, preferably driven by a central architecture description to ensure consistency across the compiler, functional, timing, and power simulators and HDL generation."

The Tensilica configuration and tool-generation process has been described in Section 1 of this chapter. It provides software tools to both map and compile application code and benchmarks to any specific configuration, and the performance-profiling capabilities in the ISS and IDE allow the quality of the results to be evaluated, and they allow the optimal fine-tuning of a configuration. The software-generation process is inherently retargetable and is driven by centralized architectural descriptions, thus ensuring the desired consistency across compilers, the instruction set simulator and the processor's HDL description. However, Tensilica does not use a single architectural description mechanism. The first-generation micro-architecture used a combination of parameterized HDL description, designer-defined configuration parameters, and the TIE instruction-extension language, to describe a configuration architecture. The second generation micro-architecture took advantage of the general capabilities of the TIE language and compiler to describe much more of the LX micro-architecture in TIE – thus moving closer to a single source description, where most of a configuration ends up as TIE source code.

More important than a single-source architectural description is a unified configuration and generation process that can generate both the complete software and simulation toolset and the hardware implementation all via a centralized single process. As described earlier, this is the mechanism Tensilica offers.

5.   Element 4: Comprehensively Exploring the Design Space

"This should be a feedback driven process, where each design point is evaluated on the appropriate metrics, using the software environment described in element 3. Efficient exploration for a large number of design points relies on quick reconfiguration of the design environment and simulators."

Initially Tensilica offered a manual design-space exploration process. Tensilica has gone one step further in automating design space-exploration and processor-configuration optimization with the XPRES^{TM} [84] tool and methodology introduced in 2004. This technology automates the process of creating processor configurations that achieve the EEMBC "full-fury" level of optimization without manual effort. The XPRES^{TM} Compiler analyzes C/C++ code and automatically generates performance-enhancing processor extensions from its code analysis. Tensilica's Xtensa$^®$ Processor Generator accepts the output from the XPRES^{TM} Compiler and automatically produces the specified processor and a compiler that recognizes the processor's architectural extensions as native instructions. The resulting compiler automatically generates object code that uses these extensions without the need to use intrinsics, for any application source code, not just the original target code analyzed by the XPRES™ compiler.

Under EEMBC's out-of-the-box benchmarking rules, results from code compiled with such a compiler and run on such a processor qualify as out-of-the-box (not optimized) results. The results of an EEMBC Consumer benchmark run using such a processor appear in Figure 8.3. This figure compares the performance of a stock Xtensa® V processor core configuration with that of an Xtensa® LX processor core that has been automatically extended using the XPRES™ compiler. (Note: Figure 8.3 compares the Xtensa® V and Xtensa® LX microprocessor cores running at 260 and 330 MHz respectively but the scores are reported on a per-MHz basis, canceling out any performance differences attributable to clock rate. Also, the standard versions of the Xtensa® V and Xtensa® LX processor cores have the same ISA and produce the same benchmark results.)

Processor Name-Clock	Tensilica Xtensa T1050 -260	Tensilica Xtensa LX 330 MHz
ConsumermarkTM	.08696	.51997
Compress JPEG	.04796	0.056
Decompress JPEG	.06837	0.083
High Pass Grey-scale filter	0.45136	8.661
RGB to CYMK Conversion	0.50251	7.572
RGB to YIQ Conversion	0.33830	6.310

Processor Name-Clock	Tensilica Xtensa T1050 -260	Tensilica Xtensa LX 330 MHz
ConsumermarkTM	1.00	5.98
Compress JPEG	1.00	1.17
Decompress JPEG	1.00	1.21
High Pass Grey-scale filter	1.00	19.19
RGB to CYMK Conversion	1.00	15.07
RGB to YIQ Conversion	1.00	18.65

Figure 8.3. Tensilica's XPRES™ Compiler can produce a tailored processor core with enhanced performance for a target application program in very little time. © EEMBC. Reproduced by permission.

The XPRES™ Compiler boosted the stock Xtensa® processor's performance by 1.17x to 18.7x, as reported under EEMBC's out-of-the-box rules. The time required to effect this performance boost was one hour. These performance results demonstrate the sort of performance gains designers might expect from automatic processor configuration. However, these results also suggest that benchmark performance comparisons become even more complicated with the introduction of configurable-processor technology and that the designers making comparisons of processors using such benchmarks must be especially careful to understand how the benchmarks tests for each processor are conducted.

The XPRES™ Compiler technology starts with a fully functional microprocessor core, Xtensa® LX, and then adds hardware to it in the form of additional execution units and corresponding machine instructions to speed processor execution for the target application. Analysis of the source application code guides the generation of processor-configuration alternatives, which are evaluated using sophisticated performance and area estimators. Using characterized and tuned estimators allows XPRES™ to evaluate millions of possible microarchitectures as part of its design-space exploration in a few minutes.

The result of this search is a set of microprocessor configurations with a range of performance/cost characteristics (cost translates into silicon area on the SoC) presented on a Pareto-style curve as shown in Figure 8.4. The development team need only pick the configuration with the right performance and cost characteristics for the target application and then submit that configuration to Tensilica's Xtensa® Processor Generator for implementation in Verilog® or VHDL.

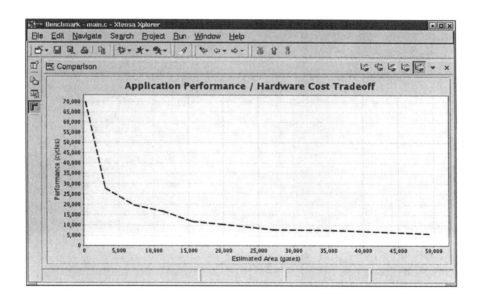

Figure 8.4. The XPRES™ Compiler presents the designer with a series of microprocessor configurations that provide increasing amounts of application-specific performance for an increasing amount of silicon area.

5.1   XPRES™ Techniques for Performance Optimization

The XPRES™ Compiler uses three techniques for creating optimized Xtensa® processor configurations: Operator fusion, SIMD (vectorization), and FLIX™ (flexible-length instruction extensions). Operator fusion notes the frequent occurrence of sequences of simple operations in program loops. The XPRES™ Compiler combines these instruction sequences into one enhanced instruction (which accelerates code execution by reducing the number of instructions executed within the loop, making the loop run faster), and by reducing the number of instructions that must be fetched from memory (thus cutting bus traffic). Figure 8.5 shows an operation dataflow graph that was generated by the XPRES™ Compiler. It has marked the operations SUB, ABS, ADD, and EXTUI as fusible.

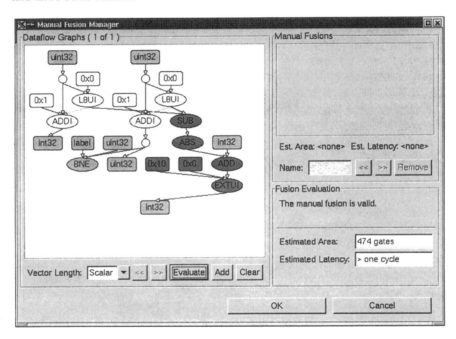

Figure 8.5.   This dataflow graph generated by the XPRES™ Compiler shows a series of operations marked as fusible. The XPRES™ Compiler estimates that a new instruction that fuses the subtraction, absolute-value, addition, and bit-field-extraction operations will require 474 additional gates.

Many loops within application programs perform the same operations on an array of data items. The XPRES™ Compiler can vectorize such loops by creating an instruction with multiple identical execution (SIMD) units that operate on multiple data items in parallel. The XPRES™ Compiler automatically tries 2-, 4-, and 8-operation SIMD vectorization in its design-space exploration. The addition of SIMD instructions to an Xtensa® processor dovetails with Tensilica's XCC C/C++ compiler, which has the ability to automatically unroll and vectorize the inner loops of application programs. The loop acceleration achieved through vectorization is usually on the order of the number of SIMD units within the enhanced instruction. Thus a 2-operation SIMD instruction approximately doubles loop performance and an 8-operation SIMD instruction speeds up loop execution by about 8x.

The third technique used by the XPRES™ Compiler to accelerate code is Tensilica's FLIX™ (flexible-length instruction extension) technology. FLIX™ instructions are multi-operation instructions like fused and SIMD instructions. However, FLIX™ instructions consist of multiple *independent* operations, in contrast with the dependent multiple operations of fused and SIMD instructions. Each operation in a FLIX™ instruction is independent of the others and the XCC C/C++ compiler can pack independent operations into a FLIX™-format instruction as needed to accelerate code. While the native Xtensa® processor instructions are 16 or 24 bits wide, FLIX™ instructions are either 32 or 64 bits wide, to allow the needed flexibility to fully describe the multiple independent operations. FLIX™ instructions can be seamlessly and modelessly intermixed with native Xtensa® instructions.

XPRES™ technology thus substantially automates design-space exploration and may be regarded as the most highly automated realization of the MESCAL methodology available in a commercial product.

6.   Element 5: Successfully Deploying the ASIP

MESCAL's element 5 – the successful deployment – can be summarized as "ASIPs that have been created to be a good match between application and architecture must be able to be programmed efficiently at a reasonable level, using programmer's models that expose critical features of the ASIP architecture and enable exploitation of unique architectural features. At the same time, application portability is an important goal."

The evolution of Tensilica's XCC compiler and XPRES™-based design-space exploration technology has gradually improved the efficiency of programming a configured Xtensa® processor. Before XPRES™, use of instruction extensions implemented via TIE was a manual process, invoked by calling intrinsics made available in C or C++ via macros defined by the processor generator. However, XPRES™ technology shares a common front end with the XCC compiler. This builds the control and dataflow graph from the application source code to both define possible instruction extensions, and to automatically infer their use when compiling the applications on the configured processor. In this way, the programmer's model has been gradually raised to a level that both exposes unique instruction extensions for manual use and allows automated inference of automatically generated instructions.

As part of their deployment of programmable platforms, some researchers have investigated the use of domain-specific languages such as Click for network applications. Tensilica has taken a complementary approach. Using Tensilica's design methodology, design flows from domain- or application-specific specification languages can be mapped into a common C or C++ intermediate form and then processed with the XPRES^{TM} Compiler, which improves application portability while still employing the ASIP approach and MESCAL philosophy. For example, code generated by the NP-Click code synthesizer (see Chapter 6) could be input to the XPRES^{TM} Compiler. This highly-automated approach requires further investigation and possible technology development.

Further consideration of ASIP deployment leads naturally to the topic of multiprocessor System-on-Chip (MPSoC) design.

6.1   Towards MPSoC Deployment

It is a rare application today that can achieve performance goals with just one ASIP, even using a configurable processor that is tailored for the target applications. As a result, building a true programmable platform entails not just building an ASIP but organizing multiple ASIPs and their interconnection network on one chip. The Tensilica economically sized configurable processor, together with its high-bandwidth interfaces encourages its use in multiprocessor SoC (MPSoC) designs. Advanced SoCs now commonly use ten or more configurable processors and some high-end SoC designs now use nearly two hundred complete processors per chip.

The choice of hardware-interconnection mechanisms among processor blocks in an SoC greatly affects performance and silicon cost and these hardware-interconnection mechanisms must directly support the interconnection requirements of multi-processor (MP) system design. Message-passing software communications have a natural correspondence to data queues, but message passing can be implemented using other types of hardware such as bus-based hardware with global memory. Similarly, the shared-memory software-communications mode has a natural correspondence to bus-based hardware, but shared-memory protocols can be physically implemented even when no globally accessible physical memory exists. This implementation flexibility allows chip designers to implement a spectrum of different task-to-task connections in ways that optimize performance, power, and cost together. Such communications-centric design approaches represent an extension of the single processor programmer's model to a new MP level.

Configurable processors offer significant flexibility in supporting arbitrated access to shared devices and memory. The basic topologies for shared memory buses are:

- *Remote global memory accessed over a general processor bus:* The processor implements a general-purpose interface that allows a wide variety of bus transactions. If the processor determines that the corresponding data is not local during a read (based on the address or due to a cache miss), the processor must make a non-local reference. The processor requests control of the bus, and when control is granted, sends the target read address over the bus. The appropriate device (for example, memory or input/output interface) decodes that address and supplies the requested data back over the bus to the processor, as shown in Figure 8.6.

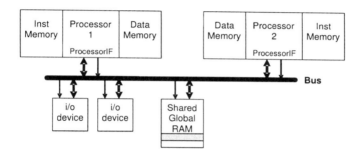

Figure 8.6. Two processors access shared memory over bus.

When two processors are communicating through global shared memory on the bus, one must acquire bus control to write the data; the other processor must later acquire bus control to read it. Each word transferred in this fashion requires two bus transactions. This approach requires modest hardware and maintains high flexibility, because the global memories and input/output interfaces are accessible over a common bus. However, the use of global memory does not scale well with the number of processors and devices, because bus traffic from multiple processors quickly overloads the bus, which leads to long and unpredictable contention latency.

- *Local processor memory accessed over a general processor bus:* Configurable processors may allow their local data memories to participate in general-purpose bus transactions. These data memories are primarily used by the processor to which they are closely coupled. However, the processor controlling the local data memory can serve as a bus slave and respond to requests on the general-purpose bus, as shown in Figure 8.7.

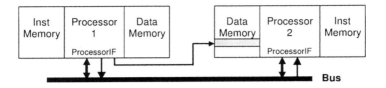

Figure 8.7. One processor accesses local data memory of a second processor over bus.

In this case, the read by Processor 1 may require access arbitration at two levels: The first when Processor 1 requests access to the general-purpose bus, and the second when the read request reaches Processor 2. The read request from Processor 1 arrives over Processor 2's bus interface and may contend with other requests for local data-memory access from tasks running on Processor 2. Two arbitration levels may increase the access latency seen by Processor 1 but Processor 2 avoids access latency almost entirely, because latency to local data memory is usually very short (one or two cycles).

This latency asymmetry between Processor 1 and Processor 2 encourages *push communication*: When Processor 1 sends data to Processor 2, it writes the data over the bus into Processor 2's local data memory. If the write is buffered, Processor 1 can continue execution without waiting for the write to complete. Thus the long latency of data transfer to Processor 2 is hidden. Processor 2 sees minimal latency when it reads the data, because the data is local. Similarly, when Processor 2 wants to send data back to Processor 1, it writes the data into Processor 1's local data memory.

- *Multi-ported local memory accessed over local bus:* When data flows in both directions between processors and latency is critical, a locally shared data memory is often the best choice for inter-task communications. Each processor uses its local data memory interface to access a shared memory, as shown in Figure 8.8. This memory could have two physical access ports (two memory references satisfied each cycle) or could have only one port controlled by a simple arbiter, where one processor's access is held off for a cycle if the other processor is using the single physical access port.

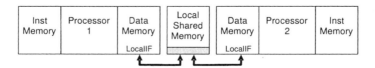

Figure 8.8. Two processors share access to local data memory.

Arbitration for a single port is preferred in area- and cost-sensitive applications, especially when shared-memory utilization is modest, because a true dual-ported memory is about twice as big per bit when compared to single-ported RAM. However, a true dual-ported memory may be the better choice when the shared memory is very small or when absolute determinism of access latency is required.

6.1.1   Direct Connect Ports.   Direct processor-to-processor connections reduce cost and latency for communication. They allow data to move directly from one processor's registers to the registers and execution units of another processor with no I/O cycles. A simple example of direct connection is shown in Figure 8.9. This example takes advantage of exportation of state registers and importation of wire values (features found in some extensible processors such as the Xtensa® LX) to create an additional dedicated interface within each processor and to directly connect them.

Whenever the Processor 1 writes a value to the output register, usually as part of some computation, that value automatically appears on the output pins of the processor. That same value is immediately available as an input value for operations in Processor 2. Wire connections can be arbitrarily wide, allowing large and non-power-of-two-sized operands to be transferred easily and quickly with very low hardware cost.

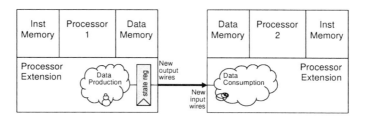

Figure 8.9. Direct processor-to-processor ports.

The operation that produces the data for the output state register may be as simple as a register-to-register transfer or it may be a complex logic function based on many other processor state values. Similarly, the input value can simply be transferred to another processor state within Processor 2 (register or memory), or it could be used as one input to a complex logic function.

6.1.2   Data Queues.   The highest-bandwidth mechanism for task-to-task communication is hardware implementation of data queues. One data queue can sustain data rates as high as one transfer every cycle or more than 10 Gbytes per second for wide operands (tens of bytes per operand at a clock rate of hundreds of MHz) because queue widths need not be tied to a processor's bus width or general-register width. The handshake between producer and consumer is implicit in the interfaces between the processors and the queue's head and tail.

When the data producer creates the data, it pushes it into the tail of the queue, assuming the queue is not full. If the queue is full, the producer stalls. When the data consumer is ready for new data, it pops it from the head of the queue, assuming the queue is not empty. If the queue is empty, the consumer stalls. The transactions are thus self-regulating, and proceed at the producer's and consumer's maximum speeds.

Queues can also be configured to provide non-blocking push and pop operations, where the producer can explicitly check for a full queue before attempting a push and the consumer can explicit check for an empty queue before attempting a pop. This mechanism allows the producer or consumer task to move to other work in lieu of stalling.

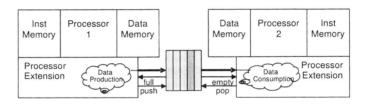

Figure 8.10.   Hardware data queue mechanism.

Some application-specific processors such as Xtensa® LX allow direct implementation of queues as part of their instruction-set extensions. An instruction can specify a queue as a destination for result values or use an incoming queue value as a source. This form of queue interface, shown in Figure 8.10, allows a new data value to be created or used each cycle on each queue interface. A complex processor extension could perform multiple queue operations per cycle, perhaps combining inputs from two input queues with local data and sending values to two output queues. The high aggregate bandwidth and low control overhead of queues allows application-specific processors to be used for applications with very high data rates where processors with conventional bus or memory interfaces are inadequate.

Queues decouple the performance of one task from another. If the rates of data production and data consumption are similar and quite uniform, the queue can be shallow. If either production or consumption rates are highly variable, a deep queue can mask this mismatch and ensure throughput at the average rate of producer and consumer, rather than at the minimum rate of the producer or the minimum rate of the consumer. Sizing the queues is an important optimization driven by good system-level simulation. If the queue is too shallow, the processor at one end of the communication channel may stall when the other processor slows for some reason. If the queue is too deep, the silicon cost will be excessive.

Because queue interfaces to processor execution units are an unusual feature of commercial microprocessor cores, it is important to discuss this interface mechanism in more depth. Queue interfaces are added to an Xtensa® LX processor through the following TIE syntax:

```
queue <queue-name>   <width> in|out
```

The name of the queue, its width, and direction are defined with the above syntax. One Xtensa® LX processor can have more than 300 queues, of variable width up to 1024 bits each. These limits are set beyond the routing limits of current silicon technology so that the processor core's architecture is not the limiting factor in the design of a system. The designer can set the real limit based on system requirements and CAD flow. Using queues, designers can trade off fast and narrow processor interfaces with slower and wider interfaces to achieve bandwidth and performance goals.

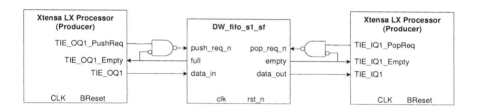

Figure 8.11. Designware synchronous FIFO used with TIE queues (The diag_n input is driven high, and the almost_full, half_full, almost_empty, and error outputs are unused).

Figure 8.11 shows how TIE queues are easily connected to simple Synopsys DesignWare® FIFO's. TIE queue push and pop requests are gated by the FIFO-empty and -full status signals to comply with the DesignWare FIFO specification. More elaborate FIFO implementations may be able to take advantage of the request signals when the FIFO is empty or full.

TIE queues serve directly as input and output operands of TIE instructions, just like a register operand, a state, or a memory interface. The following TIE syntax creates a new instruction that accumulates values from an input queue into a register file.

```
operation QACC {inout AR ACC} {in IQ1} {
    assign ACC = ACC + IQ1;
}
```

Figure 8.12 shows how TIE queues can be used just like other instruction operands in an Xtensa® LX processor. The figure also illustrates one of the key differences between queue interfaces and memory interfaces: The width of each queue interface port can be customized to the exact width desired, whether wider or narrower than the processor's standard memory interface ports.

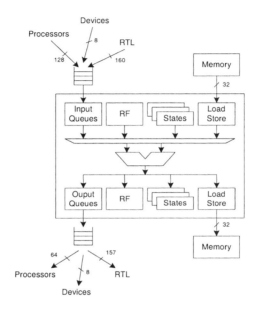

Figure 8.12. TIE queues used as instruction operands.

Whereas memory accesses often exploit temporal locality, queue data by its very nature is quite transient. In general, this pattern of reference allows queue storage to be smaller than a general-purpose memory buffer used for similar purposes. The Xtensa® LX processor itself includes 2-entry buffering for every TIE queue defined by the system designer. The area consumed by a queue's 2-entry buffer is substantially smaller than a load/store unit, which can have large combinational blocks for alignment, rotation, and sign extension of data, as well as cache-line buffers, write buffers, and complicated state machines. The processor area consumed by TIE queue interface ports is all under the designer's direct control, and can be as small or large as necessary.

6.2   Flow-Through Processing

The availability of ports and queues tied directly to a configurable processor's execution units permits the use of processors in an application domain previously reserved for hand-coded RTL logic blocks: Flow-through processing. By combining input and output queue interfaces with designer-defined execution units, it is possible to create a firmware-controlled processing block within a processor that reads values from input queues, performs a computation on those values, and outputs the result of that computation with a pipelined throughput of one clock per complete input-compute-output cycle. Figure 8.13 illustrates a simple design of such a system with two 256-bit input queues, one 256-bit output queue, and a 256-bit adder/multiplexer execution unit. Although this processor extension runs under the control of firmware, its operation bypasses the processor's memory buses and load/store unit to achieve hardware-like processing speeds.

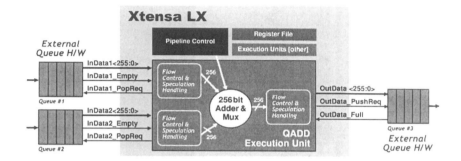

Figure 8.13. Combining queues with execution unit to add flow-through processing to a configurable processor core.

Even though there is substantial hardware in this processor extension, its definition consumes only four lines of TIE code:

```
queue InData1 256 in
queue InData2 256 in
queue OutData 256 out
operation QADD {} { in InData1, in InData2, in SumCtrl, out OutData}
    { assign OutData = SumCtrl ? (InData1 + InData2) : InData1; }
```

The first three lines of this code define the 256-bit input and output queues and the fourth line define a new processor instruction, QADD, which performs 256-bit additions or passes 256-bit data from input to output. Defining the instruction in TIE tells the Xtensa® Processor Generator to automatically add the appropriate hardware to the processor and to add the new instruction to the processor's software-development tool set.

Exploring the MP inter-processor communications space opened up by these new mechanisms is one required element of the new MPSoC programmer's model.

7.   Conclusion: Further MPSoC Progress

The MESCAL methodology and the Tensilica configurable-processor-core design approach agree on the importance of processors as the key flexible building block for SoC designs, making it possible to leverage the very high transistor count and excellent connectivity made available by nanometer silicon lithography with relatively little manual design effort. Configurable processor cores or ASIPs can achieve much higher performance than conventional, fixed-ISA processors through the addition of custom-tailored execution units, registers and register files, and specialized communication interface ports.

The disciplined MESCAL five-element methodology is easily supported by the capabilities of Tensilica's design approach. New technologies such as automated processor design-space exploration help to automate elements of the process that would typically rely on manual application. Tensilica's capabilities for heterogeneous inter-processor communication mechanisms demonstrate how the MESCAL methodology can be extended to include the creation of increasingly complex application-oriented MP subsystems.

Such a methodology could include more abstract application-task partitioning and communication programmer's models, which can be flexibly mapped onto heterogeneous sets of MP communications resources including combinations of direct connections, FIFO queues, shared memory, and bus-based architectures. If a designer starts with application code that utilizes communications abstractions, he or she can then map the high level "virtual channels" onto a variety of physical implementation forms. The use of simulation, and cost estimators to analyze the various alternatives, creates an automated "communications-space exploration" design flow which shifts attention from individual processors to the MP subsystem as a whole.

Of course, applications implemented with an MPSoC subsystem can then drive automated processor optimization and system balancing (both processing and communications) using technologies such as the XPRES™ compiler. The combination of an abstract communications model at the application level, automated mapping, processor and communications implementation generation, and design-space exploration, represents a new level in MPSoC system design automation, and elevates the MESCAL methodology to a new level.

Acronyms

ADL   Architecture Description Language. See the discussion of related work in Section 1, Chapter 4 or Section 3 in Chapter 7.

AMBA™   Advanced Microprocessor Bus Architecture by ARM Ltd.

API   Application Programming Interface. An API defines calling conventions by which an application program accesses services from the Operating System (OS) and other utilities. An API thus abstracts from the implementation of the OS kernel (or other privileged utilities) to ensure the portability of the application code.

ASIC   Application-Specific Integrated Circuit.

ASIP   Application-Specific Instruction (set) Processor.

ATM   Asynchronous Transfer Mode is a connection-oriented switching technology that organizes data into 53-byte cell units.

AVF   Architects View Framework, see Chapter 7.

BDD   Binary Decision Diagram. A BDD is a data structure that is used to represent a Boolean function. BDDs are often visualized as a rooted, directed, acyclic graph. Each non-leaf vertex is labeled by a variable and has two child nodes, representing true and false assignments. Leaves are either labeled true or false.

BGP   Border Gateway Protocol is used for exchanging routing information between gateway hosts. The routing table contains a list of known routers, the addresses they can reach, and a cost metric associated with the path to each router so that the best available route is chosen.

CAM   Content Addressable Memory. Unlike RAMs, given a data word a CAM returns the address of the data if it is stored in the CAM.

CGD   Code Generator Description.

COMA   Cache-Only Memory Architecture.

CRC   Cyclic Redundancy Check. A method for detecting transmission errors.

CSIX   Common Switch Interface. An interface standard to connect line cards with switch fabrics.

DAG   Directed Acyclic Graph.

DDR   Double Data Rate. A technology that uses both edges of the clock to transmit data, e.g. used by memories, see JEDEC JESD79D.

DES   Data Encryption Standard.

DHCP   Dynamic Host Configuration Protocol is a communication protocol that automates the assignment of IP addresses in an organization's network.

DiffServ Differentiated Services. Relative QoS scheme for IP networks, see [24].

DRR   Deficit Round Robin. Scheduling technique.

DSCP   DiffServ CodePoint. Field in packet header used by DiffServ PHB classification.

DSE   Design Space Exploration. Generic term for the process of evaluating several designs in order to determine design alternatives, see Chapter 5.

DSM   Deep-Submicron Effects. New electrical design challenges due to deep sub-micron geometries, such as interconnect delay and signal integrity.

DUT   Design Under Test.

FLIX^{TM} Flexible Length Instruction set encoding used by Tensilica (see Chapter 8).

FMO   FindMinimalOperations. Method in the Tipi framework for extracting all primitive operations from the description of the data path, see Section3.1 in Chapter 4.

FPGA   Field Programmable Gate Array.

FSM   Finite State Machine. Model of computation.

HDL   Hardware Description Language.

HLL   High Level Language.

IDE   Integrated Development Environment.

IETF   Internet Engineering Task Force *http://www.ietf.org*.

IF   Interface.

ILP   Instruction Level Parallelism.

ILP   Integer Linear Programming. Optimization technique. A system of linear equalities and linear inequalities over a set of unknown integer variables, along with a linear objective function to be maximized, has to be solved. If only some of the unknown variables are required to be integers, the problem is called mixed integer programming.

IMC	Interface Method Call.
ISA	Instruction Set Architecture. Systematic definition of an instruction set and its encoding that provides an abstraction of the underlying architecture to the programmer.
ISO	International Organization for Standardization *http://www.iso.org*.
ISS	Instruction Set Simulator.
ITRS	International Technology Roadmap for Semiconductors *http://public.itrs.net*.
ITU	International Telecommunication Union *http://www.itu.int*.
JEDEC	Joint Electron Device Engineering Council *http://www.jedec.org/*.
JIT-CCS	Just-in-Time Cache Compiled Simulation, see also Chapter 7.
KPN	Kahn Process Network. Model of computation.
LISA	Language for Instruction Set Architectures, see Chapter 7.
LPM	Longest Prefix Match. Algorithm to perform IPv4 address lookup.
MAC	Media Access Control. MAC is the lower sublayer of the OSI data link layer, defining the interface between the logical link control and the physical layer. The MAC sublayer is primarily concerned with breaking data up into data frames, transmitting frames, and controlling access to the medium.
MESCAL	Modern Embedded Systems: Compilers, Architectures, and Languages.
MII	Media Independent Interface.
MoC	Model of Computation. A MoC defines a set of rules that govern the interaction of components in the model. A MoC is executable. Useful MoCs for designing heterogeneous systems express concurrency and time.
MOPS	Million Operations Per Second.
MPEG	The Moving Picture Experts Group develops standards for video and audio compression.
MPLS	Multi-Protocol Label Switching. MPLS sets up a specific path for a given sequence of packets. The sequence is identified by a label put in each packet. Costly address lookups are avoided in a router.
MPSoC	Multi-Processor System-on-Chip.
MSI	Modified-Shared-Invalid. Cache/Memory coherence protocol.
NAPT	Network Address Port Translation. NAPT involves the mapping of port numbers, allowing multiple machines to share a single IP address.

NIC   Network Interface Card.
NoC   Network on Chip.
NPE   Network Processing Engine by Intel. Customized programmable accelerators to support particular network interfaces.
NPF   Network Processing Forum *http://www.npforum.org/*.
NPU   Network Processing Unit.
NRE costs Non-Recurring Engineering costs.
NW   Network.
OC-x   Optical Carrier levels are a set of signal rate multiples for transmitting signals on optical fiber. The base rate (OC-1) is 51.84 Mbps. Certain multiples of the base rate are used.
OSCI   Open SystemC^{TM} Initiative *http://www.systemc.org*.
OSI   Open Systems Interconnection reference model of ISO.
PCI   Peripheral Component Interconnect.
PDF   Probability Distribution Function.
PE   Processing Element.
PHB   Per Hop Behavior. Defines relative QoS level in DiffServ, such as expedited, assured, or best effort forwarding.
PHY   Physical layer in the OSI reference model.
PIF   Processor bus Interface (see Chapter 8).
POS   Packet over SONET. A communication protocol to transfer IP traffic over optical fiber.
POTS   Plain Old Telephony Service interface type.
QDR^{TM} Quadruple Data Rate *http://www.qdrsram.com*. Bus technique where separate buses are used for read and write accesses. Each of these buses uses DDR technology.
QoS   Quality of Service.
RED   Random Early Detection. Congestion avoidance technique for managing queues by randomly discarding packets at arrival.
RFC   Request For Comments published by IETF.
RTL   Register Transfer Level.
RTOS   Real-Time Operating System.
SAR   Segmentation and Reassembly.
SAT   The Boolean satisfiability problem (SAT) is defined by a Boolean expression written using AND, OR, and NOT operations. The question is: Given the expression, is there some assignment of Boolean values to the variables that will make the expression true? SAT is NP-complete.

SDP   Serial Data Processor by Motorola/Freescale, Inc. Micro-programmable engine that handles physical layer tasks of network interfaces.

SIMD   Singe Instruction Multiple Data processing principle.

SLA   Service Level Agreement. Contract between a network service provider and a customer that specifies, usually in measurable terms, what services the provider will offer to the customer.

SoC   System on Chip.

SONET   SONET is the American National Standards Institute standard for synchronous data transmission on optical media. The corresponding international standard is synchronous digital hierarchy (SDH).

SPEC   Standard Performance Evaluation Corp. *http://www.spec.org*.

SPI   System Packet Interface, published by the Optical Internetworking Forum *http://www.oiforum.com*.

TIE   Tensilica Instruction Extension language (see Chapter 8).

TLM   Transaction Level Modeling.

TOS   Type of Service. IP header field. Reused as DiffServ's DSCP field.

TTL   Time-To-Live. IP header field.

UART   Universal Asynchronous Receive / Transmit interface.

UMTS   Universal Mobile Telecommunications Service is a third-generation (3G) broadband packet-based transmission service for voice, video, and multimedia data in general.

USB   Universal Serial Bus.

VCD   Value Change Diagram.

VLIW   Very Long Instruction Word.

VOQ   Virtual Output Queuing. Queuing discipline for organizing network switches (as opposed to input queuing).

VPN   Virtual Private Network. VPN traffic is carried on public networking infrastructure. VPNs use cryptographic protocols to provide the necessary confidentiality.

WCET   Worst Case Execution Time.

XLMI   Tensilica Local Memory Interface (see Chapter 8).

ZBT®   Zero Bus Turnaround. RAM operation mode that delays the submission of write data so that read and write accesses can be better pipelined.

References

[1] S. G. Abraham, B. R. Rau, and R. Schreiber. Fast design space exploration through validity and quality filtering of subsystem designs. Technical Report HPL-2000-98, Hewlett-Packard Laboratories, July 2000.

[2] A. Agarwal, M. Horowitz, and J. Hennessy. An analytical cache model. *ACM Transactions on Computer Systems*, 7(2):184–215, 1989.

[3] A. Aharon, D. Goodman, M. Levinger, Y. Lichtenstein, Y. Malka, C. Metzger, M. Molche, and G. Shurek. Test program generation for functional verification of PowerPC processors in IBM. In *Design Automation Conference (DAC)*, 1995.

[4] I. Ahmad, M. Dhodhi, and C. Chen. Integrated scheduling, allocation and module selection for design-space exploration in high-level synthesis. *IEE Proceedings - Computers and Digital Techniques*, 142(1):65–71, Jan. 1995.

[5] G. Araujo and S. Malik. Optimal code generation for embedded memory non-homogeneous register architectures. In *8th International Symposium on System Synthesis (ISSS)*, pages 36–41, 1995.

[6] Arteris SA. NoC IP and NoC tools. *http://www.arteris.com*.

[7] G. Ascia, V. Catania, and M. Palesi. Design space exploration methodologies for IP-based system-on-a-chip. In *IEEE International Symposium on Circuits and Systems*, volume 2, pages 364–367, May 2002.

[8] G. Ascia, V. Catania, and M. Palesi. A framework for design space exploration of parameterized VLSI systems. In *ASP-DAC/VLSI Design 2002*, pages 245–250, Jan. 2002.

[9] Associated Compiler Experts. *The SuperTest Compiler Test and Validation Suite - Datasheet*.

[10] K. Atasu, L. Pozzi, and P. Ienne. Automatic application-specific instruction-set extensions under microarchitectural constraints. In *Design Automation Conference (DAC)*, 2003.

[11] S. Audenaert and P. Chandra. Network processors benchmark framework. NPF Benchmarking Workgroup, *http://www.npforum.org/*.

[12] M. Auguin, L. Capella, F. Cuesta, and E. Gresset. CODEF: a system level design space exploration tool. In *2001 IEEE International Conference on Acoustics, Speech, and Signal Processing*, volume 2, pages 1145–1148, 2001.

[13] T. Austin, E. Larson, and D. Ernst. SimpleScalar: an infrastructure for computer system modeling. *IEEE Computer*, 35(2):59–67, Feb. 2002.

[14] J. Axelsson. Architecture synthesis and partitioning of real-time systems: a comparison of three heuristic search strategies. In *5th Int. Workshop on Hardware/Software Codesign (CODES/CASHE)*, pages 161–165, Mar. 1997.

[15] A. Baghdadi, N. Zergainoh, W. Cesario, T. Roudier, and A. Jerraya. Design space exploration for hardware/software codesign of multiprocessor systems. In *11th International Workshop on Rapid System Prototyping (RSP)*, pages 8–13, 2000.

[16] F. Baker. Requirements for IP version 4 routers. Request for Comments 1812, Internet Engineering Task Force (IETF), June 1995.

[17] F. Balarin, M. Chiodo, P. Giusto, H. Hsieh, A. Jurecska, L. Lavagno, C. Passerone, A. Sangiovanni-Vincentelli, E. Sentovich, K. Suzuki, and B. Tabbara. *Hardware-Software Co-Design of Embedded Systems: The Polis Approach*. Number 404 in International Series in Engineering and Computer Science. Kluwer Academic Publishers, 1997.

[18] F. Balarin, Y. Watanabe, H. Hsieh, L. Lavagno, C. Paserone, and A. Sangiovanni-Vincentelli. Metropolis: an integrated electronic system design environment. *IEEE Computer*, 36(4):45–52, Apr. 2003.

[19] S. Bashford, U. Bieker, B. Harking, R. Leupers, P. Marwedel, A. Neumann, and D. Voggenauer. *The MIMOLA Language - Version 4.1, Reference Manual*. Computer Science Dpt., University of Dortmund, Sept. 1994. *http://ls12-www.cs.uni-dortmund.de/research/mimola-4.1.ps.gz*.

[20] A. Bender. MILP based task mapping for heterogeneous multiprocessor systems. In *European Design Automation Conference*, pages 190–197, 1996.

[21] L. Benini, D. Bertozzi, D. Bruni, N. Drago, F. Fummi, and M. Poncino. SystemC cosimulation and emulation of multiprocessor SoC designs. *IEEE Computer*, 36(4):53–59, Apr. 2003.

[22] L. Benini and G. DeMicheli. Networks on chips: A new SoC paradigm. *IEEE Computer*, 35(1):70–78, Jan. 2002.

[23] G. Berry and G. Gonthier. The Esterel synchronous programming language: Design, semantics, implementation. *Science of Computer Programming*, 19(2):87–152, Nov. 1992.

[24] S. Blake, D. L. Black, M. A. Carlson, E. Davies, Z. Wang, and W. Weiss. An architecture for differentiated services. Request for Comments 2475, Internet Engineering Task Force (IETF), Dec. 1998.

[25] T. Blickle, J. Teich, and L. Thiele. System-level synthesis using evolutionary algorithms. *Design Automation for Embedded Systems, Kluwer Academic Publishers*, 3(1):23–58, Jan. 1998.

[26] H. Blume, H. Feldkämper, and T. Noll. Model-based exploration of the design space for heterogeneous systems-on-chip. *Journal of VLSI-Signal Processing Systems*, 40(1):19–34, May 2005.

[27] H. Blume, H. Huebert, H. Feldkämper, and T. Noll. Model based exploration of the design space for heterogeneous systems on chip. In *13th Int. Conference on Application-specific Systems, Architectures and Processors (ASAP)*, pages 29–40, July 2002.

[28] S. Blythe and R. Walker. Toward a practical methodology for completely characterizing the optimal design space. In *9th International Symposium on System Synthesis*, pages 8–13, Nov. 1996.

[29] S. Blythe and R. Walker. Efficiently searching the optimal design space. In *Ninth Great Lakes Symposium on VLSI*, pages 192–195. IEEE Computer Soc., Mar. 1999.

[30] S. Blythe and R. Walker. Efficient optimal design space characterization methodologies. *ACM Transactions on Design Automation of Electronic Systems*, 5(3):322–336, July 2000.

[31] S. Bradner. Benchmarking terminology for network interconnection devices. Request for Comments 1242, Internet Engineering Task Force (IETF), July 1991.

[32] S. Bradner and J. McQuaid. A benchmarking methodology for network interconnect devices. Request for Comments 2544, Internet Engineering Task Force (IETF), Mar. 1999.

[33] G. Braun, A. Nohl, W. Sheng, J. Ceng, M. Hohenauer, H. Scharwächter, R. Leupers, and H. Meyr. A novel approach for flexible and consistent ADL-driven ASIP design. In *Design Automation Conference (DAC)*, 2004.

[34] Brecis Communications. MSP5000 multi-service processor. product brief, 2002.

[35] D. Bruni and A. B. L. Benini. Statistical design space exploration for application-specific unit synthesis. In *38th Design Automation Conference (DAC)*, pages 641–646, 2001.

[36] S. Buch. Application specific processors in industry SoC designs. In *4th Int. Seminar on Application-Specific Multi-Processor SoC (MPSoC)*, TIMA Laboratory, France, July 2004. *http://tima.imag.fr/mpsoc/2004/slides/Buch.pdf*.

[37] J. R. Burch and D. L. Dill. Automatic verification of pipeline micro-processor control. In *6th Int. Conf. on Computer Aided Verification (CAV)*, volume 818 of *Lecture Notes in Computer Science*, pages 68–80. Springer Verlag, 1994.

[38] L. Cai, D. Gajski, and M. Olivarez. Introduction of system level architecture exploration using the SpecC methodology. In *IEEE International Symposium on Circuits and Systems*, volume 5, pages 9–12, May 2001.

[39] L. Cai, S. Verma, and D. D. Gajski. Comparison of SpecC and SystemC languages for system design. Technical Report CECS 03-11, University of California, Irvine, May 2003.

[40] J. Carter, W. Hsieh, L. Stoller, M. Swanson, L. Zhang, E. Brunvand, A. Davis, C.-C. Kuo, R. Kuramkote, M. Parker, L. Schaelicke, and T. Tateyama. Impulse: building a smarter memory controller. In *5th International Symposium on High- Performance Computer Architecture (HPCA)*, pages 70–79, 1999.

[41] J. Ceng, M. Hohenauer, G. Braun, R. Leupers, G. Ascheid, and H. Meyr. C compiler retargeting based on instruction semantics models. In *Design Automation & Test in Europe (DATE)*, 2005.

[42] W. Cesario, A. Baghdadi, L. Gauthier, D. Lyonnard, G. Nicolescu, Y. Paviot, S. Yoo, A. Jerraya, and M. Diaz-Nava. Component-based design approach for multicore SoCs. In *Design Automation Conference (DAC)*, 2002.

[43] D. Chai and A. Kuehlmann. A fast pseudo-boolean constraint solver. In *Design Automation Conference (DAC)*, pages 830–835, 2003.

[44] S. Chakraborty, S. Künzli, and L. Thiele. A general framework for analysing system properties in platform-based embedded system design. In *Design, Automation and Test in Europe (DATE)*, Munich, Germany, Mar. 2003.

[45] P. Chandra, F. Hady, R. Yavatkar, T. Bock, M. Cabot, and P. Mathew. Benchmarking network processors. In P. Crowley, M. Franklin, H. Hadimioglu, and P. Onufryk, editors, *Network Processor Design: Issues and Practices*, volume 1, pages 11–25. Morgan Kaufmann Publishers, Oct. 2002.

[46] K. Chatha and R. Vemuri. An iterative algorithm for hardware-software partitioning, hardware design space exploration and scheduling. *Design Automation for Embedded Systems, Kluwer Academic Publishers*, 5(3-4):281–293, Aug. 2000.

[47] S. Chaudhuri, S. Blythe, and R. Walker. A solution methodology for exact design space exploration in a three-dimensional design space. *IEEE Transactions on Very Large Scale Integration (VLSI) Systems*, 5(1):69–81, Mar. 1997.

[48] C. Chekuri. *Approximation Algorithms for Scheduling Problems*. PhD thesis, CS-TR-98-1611, Computer Science Department, Stanford University, Aug. 1998.

[49] B. Chen and R. Morris. Flexible control of parallelism in a multiprocessor pc router. In *USENIX Annual Technical Conference*, pages 333–346, June 2001.

[50] D. Chinnery and K. Keutzer. *Closing the Gap Between ASIC and Custom: Tools and Techniques for High-Performance ASIC Design*. Kluwer Academic Publishers, 2002.

[51] C. K. Chow. Determining the optimum capacity of a cache memory. *IBM Technical Disclosure Bulletin*, 17(10):3163–3166, 1975.

[52] Cisco Systems. Technology of edge aggregation: Cisco 1000 series edge services router. product datasheet, Mar. 2001.

[53] N. Clark, H. Zhong, and S. Mahlke. Processor acceleration through automated instruction set customisation. In *36th Annual International Symposium on Microarchitecture*, Dec. 2003.

[54] E. G. Coffman and P. J. Denning. *Operating System Theory*. Prentice-Hall, 1973.

[55] P. Crowley, M. E. Fiuczynski, J.-L. Baer, and B. N. Bershad. Characterizing processor architectures for programmable network interfaces. In *International Conference on Supercomputing (ICS) 2000*, pages 54–65. ACM, May 2000.

[56] K. Crozier. A C-based programming language for multiprocessor network SoC architectures. In M. Franklin, P. Crowley, H. Hadimioglu, and P. Onufryk, editors, *Network Processor Design: Issues and Practices*, volume 2. Morgan Kaufmann, Nov. 2003.

[57] B. De Smedt and G. Gielen. WATSON: a multi-objective design space exploration tool for analog and RF IC design. In *IEEE 2002 Custom Integrated Circuits Conference*, pages 31–34, 2002.

[58] R. P. Dick and N. K. Jha. MOCSYN: Multiobjective core-based single-chip system synthesis. In *Design, Automation and Test in Europe Conference (DATE)*, pages 263–270, 1999.

[59] R. Dutta, J. Roy, and R. Vemuri. Distributed design space exploration for high-level synthesis systems. In *29th Design Automation Conference (DAC)*, pages 644–650, June 1992.

[60] S. Edwards. Compiling Esterel into sequential code. In *37th Design Automation Conference (DAC)*, pages 322–327, June 2000.

[61] EEMBC. Embedded Microprocessor Benchmark Consortium, *http://www.eembc.org/*.

[62] J. Eker, J. Janneck, E. Lee, J. Liu, X. Liu, J. Ludvig, S. Neuendorffer, S. Sachs, and Y. Xiong. Taming heterogeneity - the Ptolemy approach. *Proceedings of the IEEE*, 91(1):127–144, Jan. 2003.

[63] I. Elhanany, K. Busch, and D. Chiou. Switch fabric interfaces. *IEEE Computer*, 36(9):106–108, Sept. 2003.

[64] C. Erbas, S. C. Erbas, and A. D. Pimentel. A multiobjective optimization model for exploring multiprocessor mappings of process networks. In *CODES/ISSS*, Oct. 2003.

[65] P. Faraboschi, G. Brown, J. A. Fisher, G. Desoli, and F. Homewood. Lx: A technology platform for customizable VLIW embedded processing. In *27th Int. Symposium on Computer Architecture (ISCA)*, pages 203–213, 2000.

[66] A. Fauth, J. Van Praet, and M. Freericks. Describing instruction set processors using nML. In *European Design and Test Conference (ED&TC)*, pages 503–507, Mar. 1995.

[67] A. Ferrari and A. Sangiovanni-Vincentelli. System design: Traditional concepts and new paradigms. In *International Conference on Computer Design (ICCD)*, pages 2–12, Oct. 1999.

[68] W. Fornaciari, D. Sciuto, C. Silvano, and V. Zaccaria. A design framework to efficiently explore energy-delay tradeoffs. In *Ninth International Symposium on Hardware/Software Codesign (CODES)*, pages 260–265, 2001.

[69] W. Fornaciari, D. Sciuto, C. Silvano, and V. Zaccaria. A sensitivity-based design space exploration methodology for embedded systems. *Design Automation for Embedded Systems, Kluwer Academic Publishers*, 7(1-2), Sept. 2002.

[70] L. Fournier. Genesys-X86: An automatic test-program generator for X86 microprocessors. IBM Haifa Research Center VLSI Internal Publication.

[71] M. A. Franklin and T. Wolf. A network processor performance and design model with benchmark parameterization. In P. Crowley, M. Franklin, H. Hadimioglu, and P. Onufryk, editors, *Network Processor Design: Issues and Practices*, volume 1, pages 117–139. Morgan Kaufmann Publishers, Oct. 2002.

[72] M. A. Franklin and T. Wolf. Power considerations in network processor design. In *Second Workshop on Network Processors at the 9th International Symposium on High Performance Computer Architecture (HPCA9)*, Feb. 2003.

[73] D. Gajski, F. Vahid, S. Narayan, and J. Gong. *Specification and Design of Embedded Systems*. Prentice Hall, 1994.

[74] D. Gajski, F. Vahid, S. Narayan, and J. Gong. System-level exploration with SpecSyn. In *35th Design and Automation Conference (DAC)*, pages 812–817, June 1998.

[75] D. Gajski, J. Zhu, R. Dömer, A.Gerstlauer, and S. Zhao. SpecC: Specification language and methodology. In *Kluwer Academic Publishers*, 2000.

[76] S. Ghosh, M. Martonosi, and S. Malik. Cache miss equations: A compiler framework for analyzing and tuning memory behaviour. *ACM Transactions on Programming Languages and Systems*, 21(4):702–746, July 1999.

[77] T. Givargis, J. Henkel, and F. Vahid. Interface and cache power exploration for core-based embedded system design. In *International Conference on Computer-Aided Design (ICCAD)*, Nov. 1999.

[78] T. Givargis, F. Vahid, and J. Henkel. System-level exploration for Pareto-optimal configurations in parameterized system-on-a-chip. *IEEE Transactions on Very Large Scale Integration (VLSI) Systems*, 10(4):416–422, Aug. 2002.

[79] T. Glökler, S. Bitterlich, and H. Meyr. ICORE: A low-power application specific instruction set processor for DVB-T acquisition and tracking. *13th IEEE Workshop on Signal Processing Systems (ASIC/SOC'2000)*, Sept. 2000.

[80] T. Glökler and H. Meyr. *Design of Energy-Efficient Application-Specific Instruction Set Processors*. Kluwer Academic Publishers, June 2004.

[81] F. Glover and M. Laguna. *Tabu Search*. Kluwer Academic Publishers, July 1997.

[82] R. Gonzalez. Xtensa: a configurable and extensible processor. *IEEE Micro*, 20(2):60–70, Mar. 2000.

[83] R. Goodall, D. Fandel, A. Allan, P. Landler, and H. R. Huff. Long-term productivity mechanisms of the semiconductor industry. In *Electrochemical Society Semiconductor Silicon 2002, 9th Edition*, May 2002.

[84] D. Goodwin and D. Petkov. Automatic generation of application specific processors. In *Conference on Compilers, Architectures and Synthesis of Embedded Systems (CASES)*, pages 137–147, 2003.

[85] M. Gries. *Algorithm-Architecture Trade-offs in Network Processor Design*. PhD thesis, Diss. ETH No. 14191, Swiss Federal Institute of Technology (ETH) Zurich, Switzerland, July 2001.

[86] M. Gries. Methods for evaluating and covering the design space during early design development. *Integration, the VLSI Journal*, 38(2):131–183, Dec. 2004.

[87] M. Gries, C. Kulkarni, C. Sauer, and K. Keutzer. Comparing analytical modeling with simulation for network processors: A case study. In *Design, Automation & Test in Europe (DATE)*, pages 256–261, Mar. 2003.

[88] M. Gries, S. Weber, and C. Brooks. The Mescal architecture development system (Tipi) tutorial. Technical Report UCB/ERL M03/40, Electronics Research Laboratory, University of California at Berkeley, October 2003.

[89] T. Grötker, S. Liao, G. Martin, and S. Swan. *System Design with SystemC*. Kluwer Academic Publishers, May 2002.

[90] P. Grun, N. Dutt, and A. Nicolau. APEX: Access pattern based memory architecture exploration. In *14th International Symposium on Systems Synthesis (ISSS)*, pages 25–32, Oct. 2001.

[91] M. Guthaus, J. Ringenberg, D. Ernst, T. Austin, T. Mudge, and R. Brown. MiBench: A free, commercially representative embedded benchmark suite. In *IEEE 4th Annual Workshop on Workload Characterization*, pages 3–14, Dec. 2001.

[92] G. Hadjiyiannis, S. Hanono, and S. Devadas. ISDL: An instruction set description language for retargetability. In *34th Design Automation Conference (DAC)*, pages 299–302, June 1997.

[93] A. Halambi, P. Grun, V. Ganesh, A. Khare, N. Dutt, and A. Nicolau. EXPRESSION: A language for architecture exploration through compiler/simulator retargetability. In *Design, Automation and Test in Europe (DATE)*, pages 485–490, 1999.

[94] M. Hartoog, J. Rowson, P. Reddy, S. Desai, D. Dunlop, E. Harcourt, and N. Khullar. Generation of software tools from processor descriptions for hardware/software codesign. In *34th Design Automation Conference (DAC)*, pages 303–306, June 1997.

[95] G. Hekstra, G. L. Hei, P. Bingley, and F. Sijstermans. TriMedia CPU64 design space exploration. In *1999 IEEE International Conference on Computer Design: VLSI in Computers and Processors*, pages 599–606, 1999.

[96] J. Hennessy and D. Patterson. *Computer Architecture: A Quantitative Approach*. Morgan Kaufmann Publishers, 3rd edition, May 2002.

[97] M. D. Hill and A. J. Smith. Evaluating associativity in CPU caches. *IEEE Transactions on Computers*, 38(12):1612–1630, Dec. 1989.

[98] A. Hoffmann, H. Meyr, and R. Leupers. *Architecture Exploration for Embedded Processors with LISA*. Kluwer Academic Publishers, 2002.

[99] A. Hoffmann, O. Schliebusch, A. Nohl, G. Braun, and H. Meyr. A methodology for the design of application specific instruction set processors (ASIP) using the machine description language LISA. In *International Conference on Computer Aided Design (ICCAD)*, San Jose, CA, Nov. 2001.

[100] M. Hohenauer, H. Scharwächter, K. Karuri, O. Wahlen, T. Kogel, R. Leupers, G. Ascheid, H. Meyr, G. Braun, and H. van Someren. A methodology and tool suite for C compiler generation from ADL processor models. In *Design, Automation & Test in Europe (DATE)*, Feb. 2004.

[101] J. Horn. Multicriterion decision making. In T. Bäck, D. Forgel, and Z. Michalewicz, editors, *Handbook of Evolutionary Computation*. Institute of Physics Publishing, Bristol, UK, 1997.

[102] Z. Huang and S.Malik. Exploiting operation level parallelism through dynamically reconfigurable datapaths. In *39th Design Automation Conference (DAC)*, June 2002.

[103] C.-T. Hwang, J.-H. Lee, and Y.-C. Hsu. A formal approach to the scheduling problem in high level synthesis. *IEEE Transactions on Computer-Aided Design of Integrated Circuits and Systems*, 10(4):464–475, Apr. 1991.

[104] IC Insights. The McClean report 2002 edition: An in-depth analysis and forecast of the integrated circuit industry, 2002.

[105] IEEE, Inc., 3 Park Avenue, New York, NY 10016-5997, USA. *IEEE Std 1149.1-2001, Standard Test Access Port and Boundary-Scan Architecture*, June 2001.

[106] Infineon Technologies. Harrier-XT network processor. product brief, 1999.

[107] Intel Corp. Intel IXA SDK ACE programming framework developer's guide, June 2001.

[108] Intel Corp. Intel IXP1200 network processor family. hardware reference manual, rev. 8, Aug. 2001.

[109] Intel Corp. IXP1200 network processor microengine C RFC1812 layer 3 forwarding example design. Application Note, Sept. 2001.

[110] Intel Corp. Intel IXP1200 network processor family: Microcode programmer's reference manual, Mar. 2002.

[111] Intel Corp. Intel IXP2800 network processor. Product Brief, 2002.

[112] Intel Corp. Intel microengine C compiler support. Reference Manual, Mar. 2002.

[113] Intel Corp. IXP1200 network processor family development tools user's guide, Mar. 2002.

[114] Intel Corp. IXP425 network processor family. product brief, 2002.

[115] B. L. Jacob, P. M. Chen, S. R. Silverman, and T. N. Mudge. An analytical model for designing memory hierarchies. *IEEE Trans. Computers*, 45(10):1180–1194, 1996.

[116] D. Jani, C. Benson, A. Dixit, and G. Martin. Functional verification of configurable embedded processors. In B. Bailey, editor, *The Functional Verification of Electronic Systems: An Overview from Various Points of View*, chapter 18. IEC Press, Feb. 2005.

[117] D. Jani, G. Ezer, and J. Kim. Long words and wide ports: Reinventing the configurable processor. In *Hot Chips 16 Symposium*, Aug. 2004.

[118] A. Jantsch and H. Tenhunen, editors. *Networks on Chip*. Kluwer Academic Publishers, 2003.

[119] Juniper Networks. M160 internet backbone router. product datasheet, Dec. 2001.

[120] G. Kahn. The semantics of a simple language for parallel programming. In *IFIP Congress 74*, pages 471–475. North-Holland Publishing Co., Aug. 1974.

[121] I. Karkowski and H. Corporaal. Design space exploration algorithm for heterogeneous multi-processor embedded system design. In *35th Design and Automation Conference (DAC)*, pages 82–87, 1998.

[122] K. Karuri, M. A. A. Faruque, S. Kraemer, R. Leupers, G. Ascheid, and H. Meyr. Fine-grained application source code profiling for ASIP design. In *Design Automation Conference (DAC)*, 2005.

[123] V. Kathail, S. Aditya, R. Schreiber, B. R. Rau, D. Cronquist, and M. Sivaraman. PICO: automatically designing custom computers. *IEEE Computer*, 35(9):39–47, Sept. 2002.

[124] M. Keating and P. Bricaud. *Reuse Methodology Manual for System-on-a-Chip Designs*. Kluwer Academic Publishers, 1998.

[125] T. Kempf, M. Dörper, R. Leupers, G. Ascheid, H. Meyr, T. Kogel, and B. Vanthournout. A modular simulation framework for spatial and temporal task mapping onto multi-processor soc platforms. In *Design, Automation & Test in Europe (DATE)*, Mar. 2005.

[126] K. Keutzer, A. Newton, J. Rabaey, and A. Sangiovanni-Vincentelli. System-level design: orthogonalization of concerns and platform-based design. *IEEE Transactions on Computer-Aided Design of Integrated Circuits and Systems*, 19(12):1523 – 1543, Dec. 2000.

[127] B. Kienhuis, E. Deprettere, K. Vissers, and P. van der Wolf. An approach for quantitative analysis of application-specific dataflow architectures. In *Application-Specific Systems, Architectures, and Processors (ASAP)*, July 1997.

[128] J. Kin, C. Lee, W. Mangione-Smith, and M. Potkonjak. Power efficient mediaprocessors: design space exploration. In *36th Design Automation Conference (DAC)*, pages 321–326, 1999.

[129] T. Kogel, M. Dörper, A. Wieferink, R. Leupers, G. Ascheid, H. Meyr, and S. Goossens. A modular simulation framework for architectural exploration of on-chip interconnection networks. In *Int. Conference on Hardware/Software Codesign and System Synthesis (CODES/ISSS)*, Oct. 2003.

[130] E. Kohler, R. Morris, B. Chen, J. Jannotti, and M. F. Kaashoek. The Click modular router. *ACM Transactions on Computer Systems*, 18(3):263–297, Aug. 2000.

[131] C. Kulkarni, M. Gries, C. Sauer, and K. Keutzer. Programming challenges in network processor deployment. In *Int. Conference on Compilers, Architecture, and Synthesis for Embedded Systems (CASES)*, pages 178–187, Oct. 2003.

[132] C. Kulkarni, D. Moolenaar, L. Nachtergaele, F. Catthoor, and H. De Man. System level energy-delay exploration for multimedia applications on embedded cores with hardware caches. *Kluwer Journal of VLSI Signal Processing*, 22(1):45–57, Aug. 1999.

[133] S. Laha, J. Patel, and R. Iyer. Accurate low-cost methods for performance evaluation of cache memory systems. *IEEE Transactions on Computers*, 37(11):1325–1336, 1988.

[134] K. Lahiri, A. Raghunathan, and S. Dey. Efficient exploration of the SoC communication architecture design space. In *IEEE/ACM International Conference on Computer Aided Design (ICCAD)*, pages 424–430, Nov. 2000.

[135] K. Lahiri, A. Raghunathan, and S. Dey. Performance analysis of systems with multi-channel communication architectures. In *Proceedings of 13th International Conference on VLSI Design*, pages 530–537, Jan. 2000.

[136] K. Lahiri, A. Raghunathan, and S. Dey. System-level performance analysis for designing on-chip communication architectures. *IEEE Transactions on Computer-Aided Design of Integrated Circuits and Systems*, 20(6):768–783, June 2001.

[137] M. Langevin, E. Cerny, J. Wilberg, and H.-T. Vierhaus. Local microcode generation in system design. In P. Marwedel and G. Goossens, editors, *Code Generation for Embedded Processors*, chapter 10, pages 171–187. Kluwer Academic Press, 1995.

[138] D. Lanneer, J. Van Praet, A. Kifli, K. Schoofs, W. Geurts, F. Thoen, and G. Goossens. CHESS: Retargetable code generation for embedded DSP processors. In P. Marwedel and G. Goossens, editors, *Code Generation for Embedded Processors*, volume 317 of *SECS*, pages 85–102. Kluwer Academic Publishers, 1995.

[139] J.-Y. Le Boudec and P. Thiran. *Network Calculus: A Theory of Deterministic Queuing Systems for the Internet*. Number 2050 in LNCS. Springer Verlag, 2001.

[140] K. Leary and W. Waddington. DSP/C: A standard high level language for DSP and numeric processing. In *Int. Conf. Acoustics, Speech and Signal Processing*, pages 1065–1068, 1990.

[141] B. Lee and L. John. NpBench; a benchmark suite for control plane and data plane applications for network processors. In *Int. Conference on Computer Design (ICCD)*, Oct. 2003.

[142] C. Lee, M. Potkonjak, and W. Mangione-Smith. MediaBench: a tool for evaluating and synthesizing multimedia and communications systems. In *Thirtieth Annual IEEE/ACM International Symposium on Microarchitecture*, pages 330–335, Dec. 1997.

[143] E. Lee. Embedded software. In M. Zelkowitz, editor, *Advances in Computers*, volume 56. Academic Press, 2002.

[144] E. Lee and A. Sangiovanni-Vincentelli. A framework for comparing models of computation. *IEEE Transactions on Computer-Aided Design of Integrated Circuits and Systems*, 17(12):1217–1229, Dec. 1998.

[145] E. A. Lee. Overview of the Ptolemy project. Technical Report UCB/ERL M03/25, University of California, Berkeley, July 2003.

[146] E. A. Lee and D. G. Messerschmitt. Synchronous data flow. *IEEE Proceedings*, Sept. 1987.

[147] S. Leibson. Using performance metrics to select microprocessor cores for IC designs. In L. Lavagno, L. Scheffer, and G. Martin, editors, *The Handbook of Electronic Design Automation for Integrated Circuits*. expected to be published by CRC Press, late 2005 / early 2006.

[148] R. Leupers. *Retargetable Code Generation for Digital Signal Processors*, chapter Instruction-Set Extraction, pages 45–83. Kluwer Academic Publishers, 1997.

[149] R. Leupers, J. Elste, and B. Landwehr. Generation of interpretive and compiled instruction set simulators. In *Asia and South Pacific Design Automation Conference (ASP-DAC)*, volume 1, pages 339–342, Jan. 1999.

[150] R. Leupers and P. Marwedel. Retargetable code generation based on structural processor descriptions. *Design Automation for Embedded Systems, Kluwer Academic Publishers*, 3(1):1–36, Jan. 1998.

[151] R. Leupers and P. Marwedel. *Retargetable Compiler Technology for Embedded Systems - Tools and Applications*. Kluwer Academic Publishers, Oct. 2001.

[152] Y.-T. S. Li, S. Malik, and A. Wolfe. Performance estimation of embedded software with instruction cache modeling. *ACM Transactions on Design Automation of Electronic Systems*, 4(3):257–279, July 1999.

[153] C. Liem and P. Paulin. Compilation techniques and tools for embedded processor architectures. In J. Staunstrup and W. Wolf, editors, *Hardware/Software Co-Design: Principles and Practice*. Kluwer Academic Publishers, 1997.

[154] P. Lieverse, P. van der Wolf, K. Vissers, and E. Deprettere. A methodology for architecture exploration of heterogeneous signal processing systems. *Kluwer Journal of VLSI Signal Processing*, 29(3):197–207, Nov. 2001.

[155] O. Lüthje. *A Methodology for Automated Analysis of Application Specific Processor Models with Respect to Test Generation*. PhD thesis, Aachen University of Technology, 2004.

[156] P. S. Magnusson, M. Christensson, J. Eskilson, D. Forsgren, and G. Hallberg. Simics: A full system simulation platform. *IEEE Computer*, 35(2):50–58, Feb. 2002.

[157] G. Martin and H. Chang, editors. *Winning the SoC Revolution*. Springer, 2003.

[158] S. McCanne, C. Leres, and V. Jacobson. Tcpdump 3.4. documentation, 1998.

[159] S. A. McKee, W. A. Wulf, J. H. Aylor, R. H. Klenke, M. H. Salinas, S. I. Hong, and D. A. Weikle. Dynamic access ordering for streamed computations. *IEEE Transactions on Computers*, 49(11), Nov. 2000.

[160] G. Memik, W. H. Mangione-Smith, and W. Hu. NetBench: A benchmarking suite for network processors. In *IEEE/ACM International Conference on Computer-Aided Design (ICCAD)*, Nov. 2001.

[161] M. Miranda, C. Ghez, C. Kulkarni, F. Catthoor, and D. Verkest. Systematic speed-power memory data-layout exploration for cache controlled embedded multimedia applications. In *International Symposium on System Synthesis*, pages 107–112, Oct. 2001.

[162] P. Mishra, N. Dutt, and A. Nicolau. Functional abstraction driven design space exploration of heterogeneous programmable architectures. In *International Symposium on System Synthesis*, pages 256–261, Oct. 2001.

[163] P. Mishra, P. Grun, N. Dutt, and A. Nicolau. Processor-memory co-exploration driven by a memory-aware architecture description language. In *14th International Conference on VLSI Design*, 2001.

[164] P. Mishra, F. Rousseau, N. Dutt, and A. Nicolau. Architecture description language driven design space exploration in the presence of co-processors. In *Tenth Workshop on Synthesis And System Integration of MIxed Technologies (SASIMI)*, Oct. 2001.

[165] S. Mohanty, V. K. Prasanna, S. Neema, and J. Davis. Rapid design space exploration of heterogeneous embedded systems using symbolic search and multi-granular simulation. In *Workshop on Languages, Compilers, and Tools for Embedded Systems (LCTES)*, June 2002.

[166] C. Monahan and F. Brewer. Scheduling and binding bounds for RT-level symbolic execution. In *International Conference on Computer Aided Design (ICCAD)*, pages 230–235, 1997.

[167] G. E. Moore. Cramming more components onto integrated circuits. *Electronics*, 38(8), Apr. 1965.

[168] Motorola/Freescale Semiconductor Inc. C-Ware™ software toolset: Product brief, 2003.

[169] F. Moya, J. Moya, and J. Lopez. Evaluation of design space exploration strategies. In *25th EUROMICRO Conference*, volume 1, pages 472–476, Sept. 1999.

[170] A. Nemirovsky. Towards characterizing network processors: Needs and challenges. XStream Logic (now Clearwater Networks), whitepaper, Nov. 2000.

[171] D. Newman. Internet core router test. Light Reading, *http://www.lightreading.com/*, Mar. 2001.

[172] A.-T. Nguyen, M. Michael, A. Sharma, and J. Torrellas. The Augmint multiprocessor simulation toolkit for Intel x86 architectures. In *International Conference on Computer Design (ICCD)*, pages 486–490, Oct. 1996.

[173] K. Nichols, S. Blake, F. Baker, and D. Black. Definition of the differentiated services field (DS field) in the IPv4 and IPv6 headers. Request for Comments 2474, Internet Engineering Task Force (IETF), Dec. 1998.

[174] J. Nickolls, L. Madar III, S. Johnson, V. Rustagi, K. Unger, and M. Choudhury. Calisto: A low-power single-chip multiprocessor communications platform. *IEEE Micro*, 23(2):29–43, Mar. 2003.

[175] A. Nohl, G. Braun, O. Schliebusch, R. Leupers, H. Meyr, and A. Hoffmann. A universal technique for fast and flexible instruction-set architecture simulation. In *Design Automation Conference (DAC)*, pages 22–27, June 2002.

[176] A. Nohl, V. Greive, G. Braun, A. Andreas, R. Leupers, O. Schliebusch, and H. Meyr. Instruction encoding synthesis for architecture exploration using hierarchical processor models. In *40th Design Automation Conference (DAC)*, pages 262–267, 2003.

[177] V. Norkin, G. C. Pflug, and A. Ruszcynski. A branch and bound method for stochastic global optimization. *Mathematical Programming*, 83:425–450, 1998.

[178] O. Ogawa, K. Shinohara, Y. Watanabe, H. Niizuma, T. Sasaki, Y. Takai, S. de Noyer, and P. Chauvet. A practical approach for bus architecture optimization at transaction level. In *Design, Automation & Test in Europe (DATE), Designers' Forum*, pages 176–181, 2003.

[179] V. S. Pai, P. Ranganathan, and S. V. Adve. RSIM: An exection-driven simulator for ILP-based shared-memory multiprocessors and uniprocessors. In *3rd International Symposium on High Performance Computer Architecture (HPCA)*, pages 72–83, Feb. 1997.

[180] M. Palesi and T. Givargis. Multi-objective design space exploration using genetic algorithms. In *Tenth International Symposium on Hardware/Software Codesign (CODES)*, pages 67–72, May 2002.

[181] V. Pareto. *Cours d'Economie Politique*. F.Rouge, Lausanne, 1896.

[182] D. A. Patterson and J. Hennessy. *Computer Organization & Design: The Hardware/Software Interface*. Morgan Kaufmann, 2 edition, 1997.

[183] P. Paulin and J. Knight. Force-directed scheduling for the behavioral synthesis of ASICs. *IEEE Transactions on Computer-Aided Design of Integrated Circuits and Systems*, 8(6):661 – 679, June 1989.

[184] P. Paulin, C. Liem, T. May, and S. Sutarwala. FlexWare: A flexible firmware development environment for embedded systems. In P. Marwedel and G. Goosens, editors, *Code Generation for Embedded Processors*. Kluwer Academic Publishers, 1995.

[185] P. Paulin, C. Pilkington, and E. Bensoudane. StepNP: a system-level exploration platform for network processors. *IEEE Design & Test of Computers*, 19(6):17–26, Nov. 2002.

[186] S. Pees, A. Hoffman, and H. Meyr. Retargetable compiled simulation of embedded processors using a machine description language. *ACM Transactions on Design Automation of Electronic Systems*, 5(4):815–834, 2000.

[187] S. Pees, A. Hoffmann, and H. Meyr. Retargeting of compiled simulators for digital signal processors using a machine description language. In *Design, Automation and Test in Europe Conference (DATE)*, pages 669–673, 2000.

[188] S. Pees, A. Hoffmann, V. Zivojnovic, and H. Meyr. LISA-machine description language for cycle-accurate models of programmable DSP architectures. In *36th Design Automation Conference (DAC)*, pages 933–938, June 1999.

[189] H. Peixoto and M. Jacome. Algorithm and architecture-level design space exploration using hierarchical data flows. In *IEEE International Conference on Applications-Specific Systems, Architectures and Processors (ASAP)*, pages 272–282, 1997.

[190] A. Pimentel, L. Hertzberger, P. Lieverse, P. van der Wolf, and E. Deprettere. Exploring embedded-systems architectures with Artemis. *IEEE Computer*, 34(11):57–63, Nov. 2001.

[191] A. Pimentel, S. Polstra, F. Terpstra, A. van Halderen, J. Coffland, and L. Hertzberger. Towards efficient design space exploration of heterogeneous embedded media systems. In *Embedded processor design challenges. Systems, architectures, modeling, and simulation - SAMOS*, volume 2268 of *LNCS*, pages 57–73. Springer-Verlag, 2002.

[192] J. L. Pino, S. Ha, E. A. Lee, and J. T. Buck. Software synthesis for DSP using Ptolemy. *Journal on VLSI Signal Processing*, 9(1):7–21, Jan. 1995.

[193] J. Plofsky. The changing economics of FPGAs, ASICs and ASSPs. *RTC Magazine*, Apr. 2003.

[194] T. Puzak. *Analysis of cache replacement algorithms*. PhD thesis, University of Massachusetts, 1985.

[195] W. Qin and S. Malik. Architecture description languages for retargetable compilation. In Y. N. Srikant and P. Shankar, editors, *The Compiler Design Handbook: Optimizations & Machine Code Generation.* CRC Press, 2002.

[196] V. Rajesh and R. Moona. Processor modeling for hardware software codesign. *Int. Conference on VLSI Design,* Jan. 1999.

[197] D. S. Rao and F. Kurdahi. Hierarchical design space exploration for a class of digital systems. *IEEE Transactions on Very Large Scale Integration (VLSI) Systems,* 1(3):282–295, Sept. 1993.

[198] M. Reshadi, N. Bansal, P. Mishra, and N. Dutt. An efficient retargetable framework for instruction-set simulation. In *Int. Conference on Hardware/Software Codesign and System Synthesis (CODES/ISSS),* Oct. 2003.

[199] K. Richter, D. Ziegenbein, M. Jersak, and R. Ernst. Bottom-up performance analysis of HW/SW platforms. In *Design and Analysis of Distributed Embedded Systems, IFIP 17th World Computer Congress - TC10 Stream on Distributed and Parallel Embedded Systems (DIPES),* volume 219 of *IFIP Conference Proceedings,* Montréal, Québec, Canada, Aug. 2002. Kluwer.

[200] E. Rijpkema, K. G. W. Goossens, A. Rădulescu, J. Dielissen, J. van Meerbergen, P. Wielage, and E. Waterlander. Trade offs in the design of a router with both guaranteed and best-effort services for networks on chip. In *Design, Automation & Test in Europe (DATE),* pages 350–355, Mar. 2003.

[201] M. A. V. Rodriguez, J. M. S. Perez, R. M. de la Montana, and F. A. Z. Gallardo. Simulation of cache memory systems on symmetric multiprocessors with educational purposes. In *International Congress in Quality and in Technical Education Innovation,* volume 3, pages 47–59, Sept. 2000.

[202] M. Rosenblum, E. Bugnion, S. Devine, and S. Herrod. Using the SimOS machine simulator to study complex computer systems. *ACM Transactions on Modeling and Computer Simulation,* 7(1):78–103, Jan. 1997.

[203] C. Rowen and S. Leibson (ed.). *Engineering the Complex SOC.* Prentice-Hall PTR, 2004.

[204] J. Rowson and A. Sangiovanni-Vincentelli. Interface-based design. In *Design Automation Conference (DAC),* 1997.

[205] S. Rubin, M. Levinger, R. R. Pratt, and W. P. Moore. Fast construction of test-program generators for digital signal processors. In *International Conference on Acoustics, Speech, and Signal Processing (ICASSP)*, 1999.

[206] J. Sato, M. Imai, T. Hakata, A. Alomary, and N. Hikichi. An integrated design environment for application specific integrated processor. In *Int. Conference on Computer Design (ICCD)*, pages 414–417, Oct. 1991.

[207] C. Sauer, M. Gries, J. Gomez, and K. Keutzer. Towards a flexible network processor interface for RapidIO, Hypertransport, and PCI-Express. In M. Franklin, P. Crowley, H. Hadimioglu, and P. Onufryk, editors, *Network Processor Design: Issues and Practices*, volume 3, pages 55–80. Morgan Kaufmann Publishers, Feb. 2005.

[208] C. Sauer, M. Gries, and S. Sonntag. Modular domain-specific implementation and exploration framework for embedded software platforms. In *Design Automation Conference (DAC)*, June 2005.

[209] O. Schliebusch, G. Ascheid, D. Kammler, A. Chattopadhyay, R. Leupers, and H. Meyr. JTAG interface and debug mechanism generation for automated ASIP design. In *GSPx*, Sept. 2004.

[210] O. Schliebusch, A. Chattopadhyay, D. Kammler, G. Ascheid, R. Leupers, H. Meyr, and T. Kogel. A framework for automated and optimized ASIP implementation supporting multiple hardware description languages. In *ASP-DAC*, Jan. 2005.

[211] O. Schliebusch, A. Chattopadhyay, M. Steinert, G. Braun, A. Nohl, R. Leupers, G. Ascheid, and H. Meyr. RTL processor synthesis for architecture exploration and implementation. In *Design, Automation & Test in Europe (DATE) - Designers Forum*, Feb. 2004.

[212] O. Schliebusch, A. Chattopadhyay, E. M. Witte, D. Kammler, G. Ascheid, R. Leupers, and H. Meyr. Optimization techniques for ADL-driven RTL processor synthesis. In *16th IEEE Int. Workshop on Rapid System Prototyping (RSP)*, June 2005.

[213] E. Schnarr, M. D. Hill, and J. R. Larus. Facile: A language and compiler for high-performance processor simulators. In *ACM SIGPLAN Conference on Programming Language Design and Implementation (PLDI)*, pages 321–331, June 2001.

[214] M. Schwiegershausen and P. Pirsch. A system level design methodology for the optimization of heterogeneous multiprocessors. In *Eighth International Symposium on System Synthesis*, pages 162–167, 1995.

[215] D. Sciuto, F. Salice, L. Pomante, and W. Fornaciari. Metrics for design space exploration of heterogeneous multiprocessor embedded systems. In *Tenth International Symposium on Hardware/Software Codesign (CODES)*, pages 55–60, 2002.

[216] M. Sgroi, M. Sheets, A. Mihal, K. Keutzer, S. Malik, J. Rabaey, and A. Sangiovanni-Vincentelli. Addressing the system-on-a-chip interconnect woes through communication-based design. In *Design Automation Conference (DAC)*, pages 667–672, June 2001.

[217] H. Shachnai and T. Tamir. Polynomial time approximation schemes for class-constrained packing problems. In *Workshop on Approximation Algorithms*, pages 238–249, 2000.

[218] N. Shah. Understanding network processors. Master's thesis, Dept. of Electrical Eng. and Computer Sciences, University of California, Berkeley, September 2001.

[219] N. Shah and K. Keutzer. Network processors: Origin of species. In *17th International Symposium on Computer and Information Sciences (ISCIS XVII)*, Oct. 2002.

[220] N. Shah, W. Plishker, and K. Keutzer. NP-Click: A programming model for the Intel IXP1200. In M. Franklin, P. Crowley, H. Hadimioglu, and P. Onufryk, editors, *Network Processor Design: Issues and Practices*, volume 2, chapter 9. Morgan Kaufmann, Nov. 2003.

[221] N. Shah, W. Plishker, K. Ravindran, and K. Keutzer. NP-Click: A productive software development approach for network processors. *IEEE Micro*, 24(5):45–54, Sept. 2004.

[222] M. Shreedhar and G. Varghese. Efficient fair queuing using Deficit Round-Robin. *IEEE/ACM Transactions on Networking*, 4(3):375–385, June 1996.

[223] G. Snider. Spacewalker: Automated design space exploration for embedded computer systems, HPL-2001-220. Technical report, HP Laboratories Palo Alto, Sept. 2001.

[224] Sonics Inc. Siliconbackplane™ III. *http://www.sonicsinc.com*.

[225] A. Srinivasan, P. Holman, J. Anderson, S. Baruah, and J. Kaur. Multiprocessor scheduling in processor-based router platforms: Issues and ideas. In M. Franklin, P. Crowley, H. Hadimioglu, and P. Onufryk, editors, *Network Processor Design: Issues and Practices*, volume 2, chapter 5. Morgan Kaufmann, Nov. 2003.

[226] V. Srinivasan, S. Radhakrishnan, and R. Vemuri. Hardware software partitioning with integrated hardware design space exploration. In *Design, Automation and Test in Europe (DATE)*, pages 28–35, Feb. 1998.

[227] M. Steinert, O. Schliebusch, and O. Zerres. Design flow for processor development using SystemC. In *SNUG Europe*, Munich, Germany, Mar. 2003.

[228] R. A. Sugumar and S. G. Abraham. Efficient simulation of caches under optimal replacement with application to miss characterization. In *ACM Sigmetrics Conference on Measurements and Modeling of Computer Systems*, pages 24–35, May 1993.

[229] R. Szymanek and K. Kuchcinski. Design space exploration in system level synthesis under memory constraints. In *25th EUROMICRO Conference*, volume 1, pages 29–36, 1999.

[230] Y. Tamir and G. Frazier. High performance multi-queue buffers for VLSI communication switches. In *15th Int. Symponsium on Computer Architecture (ISCA)*, 1988.

[231] H. Theiling, C. Ferdinand, and R. Wilhelm. Fast and precise WCET prediction by separate cache and path analyses. *Real-Time Systems, Kluwer*, 18(2-3):157–179, May 2000.

[232] L. Thiele, S. Chakraborty, M. Gries, and S. Künzli. Design space exploration of network processor architectures. In P. Crowley, M. Franklin, H. Hadimioglu, and P. Onufryk, editors, *Network Processor Design: Issues and Practices*, volume 1, pages 55–89. Morgan Kaufmann Publishers, Oct. 2002.

[233] H. Tomiyama, A. Halambi, P. Grun, N. Dutt, and A. Nicolau. Architecture description languages for Systems-on-Chip design. In *6th Asia Pacific Conference on Chip Design Language (APCHDL)*, pages 109–116, Oct. 1999.

[234] D. Tullsen, S. Eggers, and H. Levy. Simultaneous multi-threading: Maximizing on-chip parallelism. In *22nd Int. Symposium on Computer Architecture (ISCA)*, pages 392–403, June 1995.

[235] R. Uhlig and T. Mudge. Trace-driven memory simulation: A survey. *ACM Computing Surveys*, 29(2):128–170, June 1997.

[236] B. Vanthournout, S. Goossens, and T. Kogel. Developing transaction-level models in SystemC. White paper, CoWare Inc., Aug. 2004.

[237] O. Wahlen, M. Hohenauer, R. Leupers, and H. Meyr. Instruction scheduler generation for retargetable compilation. *IEEE Design & Test of Computers*, Jan. 2003.

[238] M. Wan, H. Zhang, V. George, M. Benes, A. Abnous, V. Prabhu, and J. Rabaey. Design methodology of a low-energy reconfigurable single-chip DSP system. *Journal of VLSI Signal Processing*, 28(1-2):47–61, May 2001.

[239] S. Weber. Mescal languages reference manual. Technical Report UCB/ERL M03/41, Electronics Research Lab, University of California at Berkeley, Oct. 2003.

[240] S. Weber, M. Moskewicz, M. Gries, C. Sauer, and K. Keutzer. Fast cycle-accurate simulation and instruction set generation for constraint-based descriptions of programmable architectures. In *Int. Conference on Hardware/Software Codesign (CODES)*, Sept. 2004.

[241] A. Wieferink, M. Dörper, T. Kogel, R. Leupers, G. Ascheid, and H. Meyr. Early ISS integration into network-on-chip designs. In *Int. Workshop on Systems, Architectures, Modeling and Simulation (SAMOS)*, July 2004.

[242] A. Wieferink, T. Kogel, G. Braun, A. Nohl, R. Leupers, G. Ascheid, and H. Meyr. A system level processor/communication co-exploration methodology for multi-processor system-on-chip platforms. In *Design, Automation & Test in Europe (DATE)*, Feb. 2004.

[243] A. Wieferink, T. Kogel, A. Nohl, A. Hoffmann, R. Leupers, and H. Meyr. A generic toolset for SoC multiprocessor debugging and synchronisation. In *IEEE Int. Conference on Application-Specific Systems, Architectures and Processors (ASAP)*, June 2003.

[244] Wind River Systems Inc. VxWorks reference manual, May 1999.

[245] T. Wolf and M. Franklin. CommBench – a telecommunications benchmark for network processors. In *2000 IEEE International Symposium on Performance Analysis of Systems and Software (ISPASS)*, pages 154–162, Apr. 2000.

[246] W. Wolf. *Modern VLSI Design: System-on-Chip Design*. Prentice Hall, 3rd edition, 2002.

[247] S. C. Woo, M. Ohara, E. Torrie, J. P. Singh, and A. Gupta. The SPLASH-2 programs: Characterization and methodological considerations. In *22nd International Symposium on Computer Architecture (ISCA)*, pages 24–36, June 1995.

[248] W. A. Wulf and S. A. McKee. Hitting the memory wall: implications of the obvious. *Computer-Architecture-News*, 23(1):20 – 24, Mar. 1995.

[249] Xilinx Corp. Virtex pro II data sheet.

[250] C. Ykman-Couvreur, J. Lambrecht, D. Verkest, F. Catthoor, A. Nikolo-giannis, and G. Konstantoulakis. System-level performance optimization of the data queueing memory management in high-speed network processors. In *39th Design Automation Conference (DAC)*, June 2002.

[251] X. Zhu and S. Malik. Using a communication architecture specification in an application-driven retargetable prototyping platform for distributed processing. In *Design Automation and Test in Europe (DATE)*, Feb. 2004.

[252] V. Zivkovic, E. Deprettere, P. van der Wolf, and E. de Kock. Design space exploration of streaming multiprocessor architectures. In *IEEE Workshop on Signal Processing Systems (SIPS)*, pages 228–234, 2002.

[253] V. D. Zivkovic, E. Deprettere, E. de Kock, and P. v. d. Wolf. Fast and accurate multiprocessor architecture exploration with symbolic programs. In *Design, Automation and Test in Europe(DATE)*, 2003.

Index